Imaging Applied to Animal Mummification in Ancient Egypt

Lidija Mary McKnight

BAR International Series 2175
2010

Published in 2016 by
BAR Publishing, Oxford

BAR International Series 2175

Imaging Applied to Animal Mummification in Ancient Egypt

ISBN 978 1 4073 0719 0

BAR Publishing is the trading name of British Archaeological Reports (Oxford) Ltd. British Archaeological Reports was first incorporated in 1974 to publish the BAR Series, International and British. In 1992 Hadrian Books Ltd became part of the BAR group. This volume was originally published by Archaeopress in conjunction with British Archaeological Reports (Oxford) Ltd / Hadrian Books Ltd, the Series principal publisher, in 2010. This present volume is published by BAR Publishing, 2016.

Printed in England

BAR
PUBLISHING

BAR titles are available from:

BAR Publishing
122 Banbury Rd, Oxford, OX2 7BP, UK
EMAIL info@barpublishing.com
PHONE +44 (0)1865 310431
FAX +44 (0)1865 316916
www.barpublishing.com

List of Contents

Please note that the CD referred to within the text has now been replaced with a download available at www.barpublishing.com/additional-downloads.html

List of Figures

List of Credits

All the photographic images included in this research project were taken by the author unless otherwise stated. Copyright for the photographs remains with the holding institution as stated in the list below.

Copyright of the radiographic images remains with the University of Manchester unless otherwise stated.

Bristol City Council
Photographs in Figures 3.1, 5.3, 5.4, 6.2, 6.5, 6.13, 6.16, 8.1, 9.1.1, 9.2.1, 9.2.2, 9.2.3, 9.2.4, 9.6.1

Cheshire West and Chester Council
Photographs in Figures 3.2, 6.19, 9.2.5. (taken by Prof. J.E. Adams).

University of Manchester
Photographs in Figures 3.4, 5.1, 5.15, 5.20, 6.1, 6.3, 6.7, 6.29 – 6.38, 9.5.1, 9.5.2

Stockport Metropolitan Borough Council
Photographs in Figure 6.40

Bolton Museum and Archive Service
Photographs in figures 3.5 (Imaging © York District Hospital), 5.12, 5.18, 6.26, 7.8, 8.2, 8.3, 2.18, 7.19

University of Liverpool
Photographs in figures 4.1 (taken by Prof. J.E. Adams), 7.1.

For Toby Alexander

and

Amy Sophia

In Loving Memory

of my father,

Timothy James Owen

5th October 1948 - 5th February 2009

Abstract

The Egyptians were the only ancient civilisation to intentionally preserve their animal dead using artificial mummification techniques. This practice formed part of a complex belief system which required that the physical remains of the body were preserved in order to ensure continuity of life in the next world. Consequently many thousands of animal mummies survive in the archaeological record, either at the numerous animal necropolises dedicated to the animal gods, or in the collections of museums and private collectors around the world.

The mummified animals from four museums in Great Britain were studied using conventional radiography (X-ray) and computed tomography (CT) in order to assess the validity and potential of the techniques. These museums are Bolton Museum and Art Gallery, Bristol Museum, The Manchester Museum and the School of Archaeology, Classics and Egyptology, University of Liverpool. In total, 127 animal mummies were transported to the Manchester Royal Infirmary and York District Hospital for radiographic analysis using modern medical equipment.

The study collection was found to contain the skeletal remains (either partial or complete) of birds, cats, dogs, fish, crocodiles, snakes and rodents, the majority of which are believed to have been manufactured as votive offerings. Alongside these mummies were twenty-eight pseudo-mummies which contained no animal skeletal material (twenty-two percent of the total collection). In addition, three unusual human-faced specimens believed to represent animal mummies were found to contain no human or animal skeletal material.

X-ray provided an excellent initial insight into the contents of the mummy bundles and should be the first method of analysis in any research project. CT was able to add significant detail and aid interpretation in a number of cases, particularly those specimens which had yielded inconclusive X-ray results or those which were found to contain unidentifiable anomalies. It appears to have limited potential in the study of small mummies as the dimensions do not allow easy differentiation between structures.

Computer manipulation of CT data using complex software allows for improved visualisation and easier interpretation of the contents of the mummy bundles. However, the dehydrated nature of mummified soft-tissue can give it a similar appearance to bone which can cause misleading results.

The radiographic examination of the mummies revealed a wealth of information on both the animals themselves and the techniques which the embalmers had used in order to preserve them. Limited evidence was found to suggest evisceration, evidence of pathological conditions or cause of death; however a great deal of information regarding the positioning of the skeleton, the form of the resulting mummy bundle and the methods of wrapping were highlighted.

In conclusion, the study identified the enormous potential of non-destructive radiographic techniques in the investigation of mummified animal material.

The Author

1997 – 2000	BSc Archaeology, University of York
2000 – 2001	MSc Biomedical and Forensic Studies in Egyptology, University of Manchester
2004 – 2008	PhD Life Sciences, University of Manchester

A lifelong interest in ancient Egypt prompted the decision to conduct a research project on the subject during my Undergraduate degree in Archaeology at the University of York (1997 – 2000). It was on the advice of my supervisor, Professor Don Brothwell, that I embarked on a quest to study mummified animals believing them to represent a poorly studied area of ancient Egyptian history. The resulting dissertation was instrumental in establishing a methodology for the radiographic study of such remains and I continued this research during my Masters degree in Biomedical and Forensic Egyptology at the University of Manchester. In September 2004, I returned to the University of Manchester to begin a PhD in Biomedical Egyptology. This research project was originally submitted as my PhD thesis.

Acknowledgements

I would like to express my heartfelt thanks to a number of people who have assisted me during this research. Firstly, I am indebted to Professors' Rosalie David and Judith Adams for their enthusiasm, support, patience and the immense amount of knowledge that they brought to the project. Special thanks go to Judith Adams for her time spent obtaining and manipulating the CT images seen in this study. Also to Dr. Keith White of the Faculty of Life Sciences, for his general support and advice throughout the process. Grateful thanks go also to Professor Don Brothwell of the Department of Archaeology, University of York, for his encouragement and advice throughout my academic career.

I would like to express my gratitude to the museums who allowed me to study their collections during this research project; namely Bolton Museum and Art Gallery, The Manchester Museum, Bristol Museum and the School of Archaeology, Classics and Egyptology, University of Liverpool. Without this support, the scope of the project would have been extremely limited. To the curators and conservators, from these institutions, I owe a huge debt of gratitude.

I would like to thank Dr. Joanne Cooper (Natural History Museum), Dr. Henry McGhie (The Manchester Museum), Wim Van Neer (Museum of Central Africa, Belgium), Mark Beech (Department of Archaeology, University of York), and staff at the Environmental Archaeology Unit (University of York) for their assistance with the identification of the animal species.

Thanks go also to the team of dedicated radiologists and technicians at York District Hospital and the Manchester Royal Infirmary where the radiography for this project was conducted. Their expertise and patience was inspirational. In particular, grateful thanks go to Chrissie Alsop, Manager of Clinical and Research Imaging Facilities, Imaging Science and Biomedical Engineering, University of Manchester, for her assistance with all aspects of the radiographic process.

On a personal note, to Martin and Peggy who supported my work whole-heartedly in life, but who will sadly not be able to witness its culmination. To my family, friends and colleagues for their continued support over the years, and more recently to my husband James and our son Toby. Thanks also to 'the bump' (Amy Sophia) for not making life too awkward during the final stages of this research and for providing the impetus to finish writing!

Chapter One
INTRODUCTION

1.0 – INTRODUCTION

Egyptology is the study of the civilisation of ancient Egypt incorporating the archaeological remains, literature and art sources – both primary (contemporary) and secondary (from later time periods). Egyptologists are tremendously fortunate that there is a wealth of material which has survived the centuries to the present day. However, this material still only represents isolated pieces in a massive jigsaw puzzle – one which presents continual challenges to those who attempt to investigate the complexities of the ancient Egyptian society.

The civilisation of the ancient Egyptians has left behind numerous lasting reminders of their technological ingenuity and amazing engineering feats which have become synonymous with what must surely be one of the worlds most interesting and awe-inspiring cultures. The sheer number and quality of these visual remains are unparalleled across both the ancient and modern world and stand testimony to the strength of the religious and political beliefs which underpinned their society. The ancient Egyptians devoted much time and considerable effort to constructing a land which fulfilled all their needs, both materially and spiritually, in life, death and the afterlife. The complexity of their civilisation and the arid environment in which they lived, has helped to ensure that much of their material world has survived to this day.

Egypt was a popular destination for eighteenth century travellers during the so-called Grand Tour when wealthy socialites took lengthy expeditions to foreign climes to visit the many wonders of the ancient world. It was as a result of this that many Egyptian antiquities found their way to Europe in the possession of these travellers. Over time, many have been donated to museums and institutions across the world as the legacy of these travellers' estates. The origins of modern Egyptology can be traced back to the 'treasure-hunting' expeditions of these early travellers, but its roots lie much farther back in time with the writings of historians such as Herodotus who described in detail the wonders of the ancient Egyptians civilisation as they witnessed them firsthand.

The study of Egyptology is an immense subject encompassing numerous sub-disciplines and areas of interest. This is testimony to the enormous amount of archaeological material which survives enabling a fragmented image of ancient Egyptian life to be formulated. Egyptologists tend to specialise in one particular area of research, but are able to draw on the expertise of fellow researchers in order to gain a better understanding of the scene as a whole. It is this multi-disciplinary approach coupled with the wealth of material which has made the study of ancient Egypt so popular and has ensured that it has a prosperous future in the academic world.

1.1 – ANCIENT EGYPTIAN BELIEFS

Egyptologists are responsible for researching an ancient civilisation and making judgements about how the individuals (both human and animal) lived and died from the evidence that is identified. It is worth pointing out that they do so, with the best intentions in the world, from a subjective standpoint. We, as modern individuals, are inherently different people to those who lived in Egypt thousands of years ago, with different views on religion, life, death and the world order.

Modern Christian beliefs teach us that death represents the end of life on earth, but that the soul lives on in heaven for eternity. The disposal of the physical remains by either burial or cremation indicates that this belief system does not require the body to be kept in any form. The temporary preservation of the body through chemical embalming is performed largely for hygiene purposes whilst maintaining the physical appearance of the person should anyone wish to view the remains before committal. It is this that separates modern beliefs from those of the ancient Egyptians, and it is this distinction which drives our fascination in them.

The belief system of the ancient Egyptians was very different to that of modern day Christians. Needless to say, they would no doubt be horrified by our actions regarding the death of our ancestors and loved ones simply because they contravene a number of their established doctrines (Germond 2001, 13-15; Owen 2001, 19). In order to be able to comprehend the motivation behind the process of intentional mummification, it is necessary to have an understanding of the religious beliefs that inspired it.

The ancient Egyptians believed that the preservation of the physical body was paramount. They believed that their earthly bodies lived on in the afterworld and were reunited with the eternal spirit. This view differs from that of reincarnation in which the soul departs the earthly body at the time of death, but that it is subsequently reborn into a different organism to begin its life-cycle once again. The ancient Egyptians essentially needed the physical body to survive the process of dying and be preserved in the likeness of a divine image. Their beliefs were formulated upon the premise that the body would be reunited with its components on the other side to live on for eternity. Mummification provided the means through which this could be achieved.

The physical body acted as a 'house' – a receptacle for the remaining components. Along with the preservation of the physical body, the ancient Egyptians believed that five further components had to be maintained to ensure the continuity of life. These were the heart, the ka, the ba, the name and the shadow. An understanding of each of these physical and non-physical components is necessary in order to fully appreciate the lengths to which they would go in order to preserve life.

The heart, rather than the brain, was viewed as the seat of a person's intellect, emotions and memory (Taylor 2001, 17). The ancient Egyptians witnessed the natural decomposition of the body, accelerated by the hot climate. They may have been aware that the internal organs began to decay immediately following the moment of death with the body essentially being destroyed from the inside out. This 'knowledge' led them to remove the internal organs during the process of mummification in an attempt to halt decomposition. The organs – the lungs, stomach, intestines and liver – were either stored outside the body in canopic jars or the organs were mummified separately before being returned to the body cavity. The heart was always left in its anatomical position within the body as it was believed that the person would require it in the afterworld.

The Book of the Dead contains many spells which discuss the heart, a number of which refer to the role of the heart in the judgement of the deceased in the hall of Osiris. Written below is an extract from Spell 30B:-

> "Oh my heart which I had from my mother! Oh my heart which I had from my mother! Oh my heart of my different ages! Do not stand up as a witness against me, do not be opposed to me in the tribunal, do not be hostile to me in the presence of the Keeper of Balance, for you are my ka which was in my body, the protector who made my members hale. Go forth to the happy place whereto we speed; do not make my name stink to the Entourage who make men. Do not tell lies about me in the presence of the gods; it is indeed well that you should hear!"
>
> (Faulkner 2000, 27-28)

The ancient Egyptians believed that every individual passed through the hall following death and that their life was judged by the gods through the answering of forty-two questions to decide whether they were to be assured an afterlife in the Field of Reeds. The heart itself was weighed against the feather of truth as a way of determining the answers to the questions so it was vitally important that it survived intact. It was customary for heart scarabs to be placed over the heart by the embalmers as insurance should the organ itself not survive the preservation process (Owen 2001, 20-21).

The ka was in effect a person's twin. It was born at the same time as the individual, but as it had no definite physical form, it was believed to reside within a statue during the lifetime of the physical individual, but at the point of death, it became a free agent with the ability to move into the corpse. Following the placement of the mummified body in the tomb, the ka had to leave the mummy, go to the offering chapel and collect victual offerings before returning to the body to provide sustenance (Taylor 2001, 18-20).

The ba essentially represented the personality of the individual. The ancient Egyptians believed that when a person died, the ba became separated from the body but was able to move about freely. They also believed that it was necessary for the ba to return to the confines of the physical body each night. For this reason, the physical body had to be preserved (Taylor 2001, 20-23).

The fourth component of each individual was the name. It was a means of expressing individuality and distinguishing one person, or animal, from another. The ancient Egyptians commonly chose names that were linked to the gods as a means of ensuring their protection. Preservation of the name was vital as it provided the means through which the individual was remembered by the living. This explains why names are included numerous times in tomb monuments, paintings and statues as a means of promoting longevity (Taylor 2001, 23-24).

The final element was the shadow. The ancient Egyptians noticed how the shadow could be seen following the individual during life and they took this to mean that it contained part of the individual, but that it was able to move outside of the physical body and return of its own accord (Taylor 2001, 24).

The ancient Egyptians had a clearly defined set of beliefs regarding the preservation of the body's components; however, it was not deemed necessary for them all to be preserved. If a couple of the components did not survive it did not matter because the remaining components worked together to ensure the survival of the individual.

The ancient Egyptians believed strongly in the myth of Osiris. The myth is fundamental to their funerary beliefs and it connects the six physical and non-physical components together to form a coherent whole. The myth also provides the motivation behind the desire to mummify the dead, both animal and human, to ensure continuity of life for eternity. There is no surviving complete copy of the ancient myth – instead Egyptologists have to rely on secondary documentation. Fragments of the myth can be seen in art and literature as early as the Old Kingdom. The most comprehensive version of the myth was recorded by the Greek priest and author Plutarch who worked in Egypt during the second century AD and made observations on his experiences (Brier 1996; Taylor 2001).

Osiris, a mythical king of Egypt, was instrumental in establishing civilised society in Egypt through the introduction of farming and animal husbandry. The Egyptians considered that Osiris was responsible for alleviating the daily pressures on the inhabitants of the Nile valley by allowing them to diversify their lifestyles. It is believed that Osiris then left Egypt in order to establish civilisation around the world (Brier 1996, 20-23).

The myth states that Osiris had an evil and jealous brother named Seth who desperately wanted to become king of Egypt. Seth formulated an elaborate plan in which he constructed a wooden chest which was designed to fit the dimensions of Osiris's body exactly. During a banquet, Seth offered to give the chest as a gift to the person who fit inside it perfectly. When Osiris's turn came to try the chest, Seth nailed down the lid and poured molten lead over the top. The chest was then deposited unceremoniously into a branch of the Nile Delta. A violent storm ensued and the chest was carried to Byblos, Lebanon, where it became trapped in the branches of a tree which grew at the side of the river. Over time the chest became increasingly entangled in the branches and eventually the tree was felled and made into one of the pillars to support a palace for the king of Byblos (Brier 1996, 20-23).

Osiris's wife, Isis, had remained in Egypt and soon learned of her husband's misfortune at the hands of Seth. She enlisted the help of the queen of Byblos to recover his body from its resting place and she returned to Egypt with the remains. Seth was enraged at the return of his brother's body, and in a fit of rage, he chopped the chest and its contents into fourteen pieces which he scattered around Egypt. Isis and her sister Nephthys (the wife of Seth) travelled across Egypt collecting Osiris's body parts. They located thirteen of the fourteen pieces; however the phallus had been discarded into the waters of the Nile and had been eaten by fish. The myth states that Isis was able to reconstruct her husband's dismembered body, including an artificial phallus, holding it together with linen bandages. By transforming herself into a bird, Isis was able to hover over Osiris's corpse and restore him to life through the chanting of magic verse (Brier 1996; Taylor 2001; Trumble 1996; Tyldesley 1999).

Essentially what Isis had created in the myth was the archetypal human mummy. She was able to bring Osiris back to life through the reassembly of the body, wrapping it in linen bands and reciting magic spells. The necessity to have a complete corpse was demonstrated through her colossal search for the body parts and the construction of an artificial phallus to complete the body. Seth, for his part, had created Egypt's first anthropoid coffin – a body-shaped protective coffin to hold the mortal remains of the deceased (Brier 1996, 20-23; Owen 2001, 22-23).

1.2 – THE THEOLOGY OF ANIMAL WORSHIP

> 'All species of animals that were buried as gods inside the animal cemeteries, made divine by the act of the Opening-of-the-Mouth ceremony, were chosen for highly theological reasons. The listed Egyptian animal world, classified into categories, was part of the continuous renewal of all creation.'
>
> (Kessler and Nureddin 2005, 127)

The ancient Egyptians could be said to have been obsessed by animals. They ruled their lives around the annual inundation of the Nile valley which brought with it the opportunity to cultivate the land and hunt the wide array of animals which were attracted to the fertile habitat. The importance of the Nile to the people of Egypt ensured that they laid great emphasis on their natural world and the environment in which they lived, worked and ultimately died. Animals played an enormous role in this eco-system.

Travellers to Egypt often looked with dismay, disbelief and occasionally disgust, on the behaviour that the native people displayed towards animals (Owen 2001, 25).

Egyptologists rely heavily on the written reports of historians such as Herodotus who travelled extensively in Egypt during the fifth century BC. Herodotus' humorous style of writing and the fact that he was writing at a time when the ancient Egyptian civilisation was beginning to decline often cause historians to look unfavourably on his work. However, it must be remembered that the Histories remain the most available comprehensive observations on the civilisation, and as such it is important that we consider them as an important firsthand account. This is especially true as there is very little surviving evidence available from the archaeological record.

Book Two of Herodotus' Histories contains information relating to animal worship and the attitude of the people of ancient Egypt towards the fauna of the country.

> 'There are not a great many wild animals in Egypt, in spite of the fact it borders Libya. Such as there are – both wild and tame – are without exception held to be

> sacred.........Anyone who deliberately kills one of these animals, is punished with death; should one be killed accidentally, the penalty is whatever the priests choose to impose; but for killing an ibis or a hawk, whether deliberately or not, the penalty is inevitably death.'
>
> (Selincourt 1985, 154-155)

The quotation above reveals the level of passion that had been bestowed on the animals of Egypt by the native people and it is also an indicator of the extreme behaviour which resulted from harming an animal. This 'passion', although obviously natural to the ancient Egyptians, seems excessive to our modern society. In fact, there can be little doubt that our fondness for animals pales into insignificance when compared to our ancient ancestors who held animals in such high regard that they were nothing less than worshipped.

The subject of animal worship in ancient Egypt is a complex one composed of many elements, some of which appear to completely contradict each other. The belief system of the Egyptians consisted of a number of contradictory factors so the fact that they viewed animals in this way is not unusual; however it is necessary to unravel all the elements and attempt to decipher what was happening. It must also be remembered that Egyptologists cannot expect to be able to understand logically an ancient belief system that simply was not intended to be logical. The ancient Egyptians had unique beliefs that must be interpreted with caution in the hope of understanding them in their original context.

The ancient Egyptians lived simple hunter-gatherer lifestyles prior to the beginnings of intentional breeding and the domestication of animals which began around the fifth millennium BC. Hunting was not only an enjoyable past-time, but also a necessary source of food. At this time, the Egyptians believed that animals were above all a food source necessary for the continuation of life in an arid land. As agriculture developed, animals became used for draught purposes, and they were also used increasingly for clothing and as household pets (Taylor 2001, 244).

Animal symbols were widely used by Predynastic communities as a means of identifying allegiance and commitment to particular population groups. It is not known whether the use of animals as symbols in this way denotes the start of a religion; however they were undoubtedly considered to be important and easily recognisable by a largely illiterate population (Owen 2001, 27; Taylor 2001, 245; Trumble 1996, 1-3).

The ancient Egyptians began to portray deities with a mixture of human and animal features during the early dynasties of their history; probably because they felt that they shared certain attributes and that the composite gods were a good way of portraying this dichotomy. It is possible this practice had its roots in an earlier religion in which the animals were the focus of worship. The practice can perhaps explain one of the most challenging aspects of animal worship: how the ancient Egyptians came to worship and hold sacred the animals that they feared the most (Houlihan 1996; 3).

An understanding of the beliefs associated with animals and the way in which they became inextricably linked to the gods themselves, provides the major motivation behind the process of animal mummification. Animals were special; whether they were feared, loathed or loved, it became necessary to protect their chances of survival in the next world. Mummification must have seemed like the logical progression.

1.3 – MUMMIFICATION IN ANCIENT EGYPT

Everything that the ancient Egyptians did was influenced by a complex belief system which essentially underpinned their whole society and way of life. It was religion which determined their everyday actions and behaviours, and the ways in which they viewed death. In essence, death was simply a part of the life-cycle which enabled a person to move on in space and time to a new life. However, death was feared intensely by the Egyptians as it symbolised the end of the physical life, so it was incredibly important to protect yourself against the possible dangers which could befall you after the actual moment of passing (Jacq 1985, xi). This concept leads us to one of the most powerful artefacts which have survived the test of time – the mummy.

Mummification in Egypt began accidentally when the hot, arid sand of the desert preserved the corpses buried in it in shallow graves. The ancient Egyptians wanted the physical body of the deceased to be preserved in a state which was as close as possible to how it had been in life, and they noticed that this type of burial seemed to ensure excellent preservation. As funerary practices evolved and tombs began to be constructed to house the remains of the deceased, it was noted that the bodies started to decompose. This move towards funerary superstructures meant that the human remains were no longer in direct contact with the naturally occurring dehydrating agents found in the sand which led to the onset of decomposition. It became necessary for the ancient Egyptians to employ methods to ensure that the body was preserved in an equally good state of preservation as the pre-Dynastic sand burials had offered, whilst still being able to fulfil their new-found desire to build elaborate tombs. In response to this dilemma, intentional mummification began (Owen 2001, 17; Taylor 2001, 47).

Initial attempts at intentional mummification were crude. The methods employed in human mummification by the ancient Egyptians can be used to create a chronology for the history of the civilisation. By examining the mummies, Egyptologists are able to determine with reasonable accuracy, the likely date of production. This chronology has been useful in dating archaeological sites where little or no literary evidence remains.

Mummies have truly provided immortality in certain respects to the ancient Egyptians who received such treatment. Their ultimate goal of the survival of the physical body has been achieved, in some cases with extremely successful results. Yet the museum collections that contain such specimens remain a priceless resource in the study of Egyptology.

Mummies are, however, not only found in human form. Immortality has also been achieved in this manner for millions of animals which were intentionally mummified by the ancient Egyptians. Thousands of animal mummies exist in museum collections worldwide, and many millions are still in their original resting places in the extensive animal necropolises which were constructed to house them for eternity. These animals are often overlooked when discussing mummification, perhaps because their physical remains are generally not accompanied by the lavish burial places which were built for the human dead, with their extensive hieroglyphs and tomb paintings which help to tell the story of the person's life. These animals are nonetheless incredibly important sources of information about the civilisation that afforded them such care and attention in both life and death.

1.4 – ANIMAL MUMMIFICATION

Prior to the unification of ancient Egypt which signalled the start of the Dynastic Period, the country was divided into a number of clans who identified themselves and each other through the use of animal imagery. Animals provided excellent symbols which could be understood by a population which was largely illiterate (Houlihan 1996; Owen 2000; Owen 2001; Taylor 2001; Trumble 1996). Also the ancient Egyptians lived closely alongside the fauna of their country and they noticed behavioural characteristics in animals which they then took to symbolise human qualities. For example, the image of a lion was often used to describe the power and strength of an influential person, usually the reigning pharoah. This theory was elaborated so that god-like qualities displayed by the deities were illustrated in animal form which led to the introduction of 'composite gods' who are depicted in Egyptian art with both human and animal features (Houlihan 1996, 2-3; Taylor 2001, 246).

Animal mummification takes on a number of forms in ancient Egypt and it is necessary to understand what particular group is being represented in the material during study so as to better grasp what the civilisation was hoping to achieve. The four categories are (Brier 1996; Ikram 2005; Owen 2001):-

Cult animals

Pets

Votive Offerings

Victual Offerings

In order to provide a context in which the animal mummies can be studied, a brief explanation of each category is now given; examples will be discussed where relevant in Chapter Two. It is necessary to define the categories here to give the animal mummies a context in which to be studied.

The Cult Animals

The cult animals have the longest history of all of the categories as the practice appears to have begun at some point during or before the fifth dynasty (Dodson 1995, 18). Much research has been conducted on these animals as they have been discovered in their original cemeteries surrounded by lavish burial paraphernalia and documentary evidence which suggest the underlying reason for their burial. They also represent national cults of which every ancient Egyptian would have knowledge, even though the cemeteries dedicated to these cults are situated at certain locations in the country – most notably at Saqqara.

Some of the most famous and well-documented cult animals appear to have been bulls who were chosen because they displayed certain physical characteristics to become the cult animal upon the death of the previous animal (Ikram 2005; Smith 1969; Pettigrew 1834). There could only be one cult animal at any one time and this animal received what can only be described as a lifestyle and treatment which rivalled that of the pharaoh. In this case it was believed that the bull was the living embodiment of the deity Ptah, and thus why there could only be a single cult animal at any one time (Davies and Friedman 1998; Owen 2000; Taylor 2001). The mothers of the sacred bulls were also revered as they had given life to the deity. At various temples around the country, one animal would be kept to represent the god but none received such lavish attention as the cult bulls.

The Pets

The next category of animal mummies is that of pet animals. Domestic pets were popular with the ancient Egyptians and it seems only natural that the owners chose to give them a 'proper burial' when they died. It is also plausible that people wanted their pets to accompany them to the afterlife, to share in what lay beyond life on this earth. The evidence that has been found to support this comes from the tombs of the pharaohs and other high-ranking officials who could afford to lavish such attention on their pets, and many are accompanied by inscriptions which tell archaeologists about their identity (Ikram 2005; Owen 2001).

The Votive Offerings

> 'Above and beyond the animal itself, stood the divinity. This should be remembered even in those few cases when the divinity behind the animal was nearly forgotten and the focus placed on the beast. The animal is only a sort of hypostatis of the god or goddess, or at the least, an intermediary between the

> divinity and the believers. The animal is sacred through delegation and association rather than on any other merits.'
>
> (Zivie and Lichtenberg 2005, 106)

Votive offerings make up the largest group by far as there are simply millions of them. During the eighteenth dynasty (c1540 – 1296 BC) there appears to have been a change in religion that saw every animal of the species being venerated instead of just one chosen one (Houlihan 1996, 7). This was continued at a low level until a revival of the cults occurred during the twenty-sixth dynasty (c664 – 525 BC). The vast majority of the votive mummies date from the later periods of Egyptian history, reaching their zenith during the Graeco-Roman and the Ptolemaic periods when foreign influences were rife in the country. They may represent a misunderstanding by these foreign rulers of the Egyptian belief in animals or simply that the worship of animals became the subject of a kind of frenzy which altered the original beliefs and caused the number of animals to be mummified to rise dramatically. This may also have been an expression of "nationalism" in the face of foreign rule (Taylor 2001, 246).

Votive mummies were manufactured as offerings to the gods. The temple complexes around the country all had associations with animals which they believed to have divine connections and pilgrims visiting these organisations wanted to offer something to the god in return for advice or protection (Owen 2001, 35; Taylor 2001, 255). The sheer volume of pilgrims wanting to partake in this caused the establishment of an industry – an early form of supply and demand in the ancient world. This industry can be analysed through study of the animal mummies it created.

Previous research into animal mummification carried out by the author has revolved around the study of these votive specimens simply because they outnumber all the other categories and they are usually the ones that have found their way into museum collections. Some of the larger institutions such as the Egyptian Museum (Cairo), the British Museum (London), the Louvre (Paris), and the Metropolitan Museum (New York) have examples of cult animals and pets in their collections, but the comparative rarity of these mummies often means that they are on public display and may not be made available for study.

Votive mummies found their way into British museums in a number of ways. The vast animal necropolises which were built to house the millions of mummies were cleared during excavations and in some cases the mummies were shipped over to Britain where they were sold by the boat-load for use as fertilizer and ballast (Trumble 1996, 42; Owen 2000, 1). This caused the destruction of a huge amount of archaeological material although cemeteries such as those at Saqqara are still complete and have been expertly excavated and recorded to provide vital information on how these votive creatures were interred (Nicholson 1994, 1).

A number of mummies which have found their way into museum collections were bought at auctions or as souvenirs by travellers to Egypt during the nineteenth and early twentieth centuries. The vast majority of these votive mummies are unprovenanced which means that there is no information as to where they came from or the period in which they were made. This must simply be viewed as one of the inherent problems with this kind of material and any research carried out can do little more than to acknowledge it.

The Victual Mummies

The fourth and final category is victual mummies – those interred in the tomb as food for the deceased person in the next world. No emphasis has been given to these 'animals' in this research project simply because the reason for their burial is a more practical one, rather than being linked to animal worship. They were not generally treated in the same way as animals that were intentionally mummified as usually they were plainly wrapped in linen, deposited on eating vessels, and preserved due to the conditions in the tomb.

There is a huge range in the quality of the finished mummy, and the care and attention taken in its production vary enormously between the categories mentioned above and at different historical periods. Cult animals have generally been eviscerated, either through a flank incision or by injecting oils into the anus to effectively dissolve the organs from within (Ikram 2005; Owen 2001; Taylor 2001; Spencer 1982). They are then treated in much the same way as human mummies, being treated with natron over a lengthy period of time, before being anointed with oils and wrapped in linen. The finishing touches made to the cult mummies are usually elaborate wrappings, painted features and ornate sarchophagi. It is likely that pets were treated in a similar way although generally not to such a great extent, primarily because they were associated with the upper levels of society who could afford to pay for such treatment. The victual mummies are normally wrapped in linen, but not treated in any way so they simply became dehydrated due to the conditions in the tomb. The votive mummies range from the quite elaborate to the extremely plain; many are simply dipped in a bitumen solution after death and wrapped in a thin layer of linen. This is no doubt an indication that the industry which created them was forced to work to capacity to supply the temples with enough offerings at the expense of the quality of workmanship.

The ancient Egyptians were the only civilisation who intentionally mummified both humans and animals, meaning that the afterlife they wanted to achieve for their human dead was also desired for animals. I believe that this is partly because they viewed all life as sacred and that humans and animals were believed to possess an equal status in life. The treatment that animals received has been viewed with scepticism by historians including Herodotus over the centuries simply because they did not understand the belief systems which caused this unique behaviour. By using biomedical techniques on the animal remains, I hope to increase our understanding of these animal cults in the same way that the human dead have been studied.

1.5 – MUMMIFICATION TECHNIQUES

The production of animal mummies reached its zenith during the Late and Graeco-Roman Periods of ancient Egyptian history, with only minimal numbers of cult, pet and victual mummies being found dating from earlier periods. This suggests that human mummification was the fore-runner of animal mummification rather than vive-versa. It seems reasonable to suggest therefore that the methods of mummification which were shown to be successful on human subjects would have been modified and used on animals.

In order to gain a clearer understanding of the treatment of animals during the mummification process, it is necessary to have an understanding of how the human dead were treated. A number of different techniques were employed in order to achieve mummification, some ultimately more successful than others, and these are discussed below.

Evisceration and Dehydration

This method is largely attributed to the Middle and New Kingdoms and was one of the most successful methods of achieving complete mummification that has been witnessed from the material remains. The method involved the removal of the internal organs via an incision which was usually made in the left side of the body, only leaving the heart in situ. The brain was removed through the nostrils to prevent putrefaction. The body was then cleaned and anointed with aromatic oils as a means of helping to combat the smell of decay, before being treated with natron which acted as a dehydrating agent absorbing moisture from the body. It is thought that the most successful examples of desiccation occurred when the natron was changed regularly allowing the full dehydrating nature of the material to be exploited. In humans, this treatment is believed to have taken around forty days to complete.

Following the desiccation process, the body was anointed once again with aromatic oils and resins before being bound in linen bandages. The viscera, which had been removed from the body, were mummified separately and either deposited in ornamental canopic jars which were stored in the burial chamber alongside the mummified corpse, or in some instances the organs appear to have been returned to the body cavities after treatment.

This method is known to have been applied to animals and in particular the cult animals such as the Apis Bulls. The length of time that the body had to be surrounded in natron would obviously depend on the size of the animal being mummified, possibly as much as fifty to sixty days for a fully-grown Apis Bull.

The site of Tuna-el-Gebel yielded the interesting discovery of an embalming room which may have been used for producing animal mummies using this technique. Evidence was found at the site to suggest that the body of the animal would have been laid on an inclining stone bed which showed deposits of resin and oils. Similar stone tables with associated pots have been excavated at the site of Memphis where the Apis Bulls were mummified (Gabra 1971, 171-172; Ikram 2005, 18).

Many examples of this mummification technique being used on animals have been discovered on archaeological sites with the majority being dated to between the Eighteenth and Twenty-first Dynasties. It is not always evident as to whether the brain has been removed, although most examples show that the internal organs are no longer present indicating evisceration. It is not always clear how the animals had been eviscerated although it is generally considered that incisions for the purpose would have been located in the abdomen or flank of the animal. Many specimens of this type show evidence that the body cavity was packed with either linen bandages or a substance such as soil or sand to provide the animal with a more recognisable body shape. This practice is also noted in human mummies of this period (Ikram 2005, 20).

Desiccation and Anointment

This method involves the evisceration of the body through an incision, and then treating the remains with natron to achieve desiccation. This appears to have taken on a number of forms which are evidenced by the archaeological remains themselves. There are indications that in many cases the dehydration process was not completed adequately enough to ensure preservation. In these cases, there is evidence to suggest that the embalmers would coat the body with generous amounts of a black tar-like substance which could be resin, oil, bitumen or pitch (Ikram 2005, 21). This coating would have prevented the growth of bacteria on the corpse, thus helping to achieve a state of preservation. Some mummies which suggest treatment of this kind have not been eviscerated which suggests that the embalmers believed that treating the body externally alone would help to preserve it.

This mummification technique appears to have been widely used in the production of animal mummies and in particular those produced as votive offerings during the Late and Graeco-Roman Periods. This would not have been as time-consuming as the evisceration and dehydration method so would have been more effective for use in a fast-paced production line environment. It is likely that the animal remains were treated with natron to some degree; however economic constraints probably facilitated that the substance was not frequently changed and that multiple animals were treated using the same batch. The level of desiccation achieved would have suffered as a result of this, leading to generally poor quality mummies.

Ikram (2005, 21) suggests that the use of hot oils and libations on bodies which had not been sufficiently desiccated would have caused charring of the remaining tissue and possibly even the combustion of the bones themselves. She suggests that the black powdery deposits found on some mummified votive offerings are all that remains of the physical body. More research into this area is undoubtedly required before definitive conclusions can be drawn.

The Use of Enemas

This technique uses an alternative method of organ removal to the traditional incision. Instead the internal organs are essentially dissolved using substances such as turpentine, oil of cedar and pine, which are inserted into the body via the anus in the form of an enema. After the substance had been introduced to the body, a bung would be inserted into the orifice to allow the dissolution of the organs. After the allotted time period, the bung would be removed allowing the liquefied organs to drain out of the body before it could be desiccated in natron. Following dehydration, the body would be anointed with oils and wrapped (Mond and Myers 1934, 62-64; Owen 2001, 29; Ikram 2005, 22).

This technique appears to have been used on both human and animal remains although the material remains demonstrating this type of treatment are relatively few in number. The Buchis Bulls and the Mother of Buchis cows found at Armant seem to have been treated in this manner judging from the archaeological discovery of enema equipment at the site. The mummified remains at this site have been destroyed by flooding so it is not possible to assess how successful the technique was on these animals; however human bodies treated in this way appear to have been well preserved (Mond and Myers 1934, 100-102).

This technique appears to have been used only on large cult animals such as the sacred bulls and is referenced in documented rituals referring to their mummification. It seems to have been common practice for the bulls to have been positioned in a sphinx-like reclining position which was achieved by dissecting the tendons in the hind legs to allow increased mobility. The animals would have been wrapped in copious amounts of linen bandages before being fixed to wooden boards using bronze or iron clamps. Elaborate decorations and gilding would have then been applied to the exterior of the mummy bundle (Moodie 1931; Mond and Myers 1934; 100-102).

Removal of the Flesh

The concept of defleshing a dead animal prior to mummification is one of considerable debate amongst Egyptologists as it raises questions regarding the reasons behind the practice and also whether or not the flesh was removed with another purpose in mind. Numerous examples of mummified animals which display evidence of defleshing have been found, most of which are attributed to the Graeco-Roman period when the production of votive offerings was at its peak (Ikram 2005, 24).

The question of whether the flesh was eaten after its removal from the corpse is largely unknown; although references to this practice have been suggested in the case of the Apis Bulls. Comparisons have been made between the 'Cannibal Hymn' taken from the Pyramid Texts and the discovery of charred bovine remains in pottery jars at the temple of Ptah. This archaeological discovery could represent the physical remains of the feast which occurred following the death of the Apis Bull in which the pharaoh consumed the meat of the animal, therefore taking on its divine powers. Dodson (2005, 77) writes:-

> 'The 'mummy' comprised a bull's head, devoid of flesh and skin, resting upon a black mass, wrapped in fine linen, and composed of broken bovine bones and pieces of gold leaf embedded in resin; this is suggestive of the bull having been at least partly consumed prior to burial'.

Alternatively, it is possible that the deceased animal was buried following death before being exhumed some time later when the skeletal remains had become disarticulated following the loss of the connective tissues. The disarticulated bones could then be anointed with ritual oils and wrapped. This practice could explain the discovery of animal mummies consisting of the wrapped disarticulated bones of species not considered to have been food-animals by the ancient Egyptians such as dogs and ibises. Ikram (2005, 24) suggests that the occurrence of such mummies could be linked to the calendar of festivals enjoyed by the animal cults and that the exhumation of buried animals perhaps formed part of these celebrations.

Molten Bitumen or Resin

This technique is witnessed mainly in the case of avian mummies created by dipping the live creatures into vats of molten bitumen or resin prior to wrapping. This was undoubtedly a fast method of killing the animal and manufacturing a mummy from the remains; however the resulting state of preservation using this technique was generally poor (Ikram 2005, 26).

Larger birds such as ibis could have been treated in this manner following death, either by immersion in vats of molten resin or through the application of the substance onto the exterior of the animal (Ikram 2005, 26). This too would have achieved poor quality mummies as the internal organs and soft tissues began to decay inside the resin casing. The absence of a dehydrating agent used in this technique would have contributed to the increased rate of decomposition.

It is possible that this method of dispatch and mummification could have been used on other small animals such as hatchling crocodiles. Their small proportions would have made them ideal candidates for this method and many display the characteristic signs of a resin or bitumen coating directly on the surface of the skin.

1.6 – POSITIONING OF THE BODY

The mummification of animals shows an enormous amount of variation throughout history and across geographic areas; however research to date has shown that certain trends are evident. The manner in which the skeletal remains were positioned following mummification largely depends on the species concerned and is obviously influenced by the size and body shape of the animal itself. Standardised body positions do occur and these are discussed in Chapter Six.

1.7 – TREATMENT FOLLOWING MUMMIFICATION

Animal mummies, like those of humans, were treated in a number of ways following the practice of mummification itself. The style of treatment given to the animal dead depends on numerous factors such as the geographic location where the object was manufactured and the period of history it belongs to. It is reasonable to suggest that there was also an element of artistic licence about these mummies with each embalmer asserting some influence over the appearance of the completed object. It is likely that cult animals were decorated in 'traditional' ways using knowledge handed down through the generations; although this expertise would have weakened as time went on without stringent recording and family associations. The decorations lavished on pet animals would almost certainly have been influenced by the wishes of the owner who would have specified how they wanted their pet to be portrayed for eternity.

There is a great deal of variation in the size and shape of animal mummies simply due to the nature of the beasts themselves. This variation differs much more widely than in human mummies where the size range was much smaller and the same basic body shape was being treated each time. For this reason, it is difficult to pinpoint common themes in animal mummy decoration; although stylistic trends do appear to have been in existence, mainly linked to the historic period.

Wrapping

The most commonly encountered method of treatment following desiccation involves the wrapping of the skeletal remains in linen bandages. Dissection of animal mummies has revealed that it was not uncommon for a creature to be treated in a succession of stages consisting either of different applications of linen bandages or an initial layer of reeds or papyrus directly covering the animal with a layer of bandages on top.

The use of linen bandages is evident from the earliest recorded animal mummies, reaching a peak in the Roman Period when more complex wrapping styles were introduced. These include the use of different coloured bandages woven together to form often elaborate patterns of squares, lines and herringbone effect. In the Graeco-Roman Period, plain bandages were used for the initial wrapping phase with the addition of facial features depicting the characteristics of the animals themselves applied to the exterior. Mummies from this period can also display appliqué designs which were intended to portray the deity to which the animal was sacrificed (Ikram 2005, 28).

Votive mummies were produced on an ancient production-line of sorts often with relatively little care and attention being lavished on the remains. The mummification process itself was generally hurried, creating poorly preserved animals simply because they were not given time to desiccate properly, if at all. The use of linen wrappings gave the embalmers a chance to essentially 'cover up' the misdemeanours of the mummification process in order to produce offerings worthy of sale to the visiting pilgrims. Therefore, the rise in the popularity of the animal cults in the Roman and Graeco-Roman Periods accompanies a general rise in the quality of the decoration and wrapping given to these animals. The more attractive the finished specimen appeared the more likely it was to sell therefore ensuring continuity of the votive business and along with it job security for the embalmers themselves.

Containers

Certain species of animal and certain animal necropolises have yielded evidence of alternative methods of treating the skeletal remains following desiccation. The site of Saqqara yielded thousands of mummified birds which had been dipped in molten resin (possibly whilst still alive) before being placed into conical ceramic pots which were then stacked floor to ceiling in the numerous chambers and corridors of the necropolis (Nicholson 1994, 9). Mummified baboons discovered at Saqqara appear to have been desiccated and placed in individual wooden boxes, into which was poured gypsum plaster which set around the mummy forming a sealed unit (Nicholson 2005, 62). The reasons behind the use of containers for housing the animal remains is largely unclear. It may represent a geographic trend as the vast majority of the animals displaying this type of treatment occur at the site of Saqqara. The reasons may have been practical, certainly in the case of the bird mummies, with the ceramic pots creating a means by which the mummified birds could be produced efficiently whilst preventing the resin-coated bodies from adhering to eachother. The shape of the pots also enabled the large volume of mummified animals to be stacked in a logical fashion to economise on the amount of space available for interment.

This method is not only found in connection with votive mummies, but occurs also in the case of pet and cult animals. Pets tend to receive lavish attention as their owners attempt to continue to care for them in the afterlife. It is not uncommon to find such animals presented in containers or sarcophagi, often manufactured to fit the animal exactly. Cult animals have been discovered in stone sarcophagi similar to pharonic burials again reinforcing their high status in life and death.

Chapter Two
THE LITERATURE

2.0 – INTRODUCTION

Animal mummies have been the subject of considerably less research than their human counterparts for numerous reasons –

1. Human remains are often seen as being more 'exciting' to Egyptologists as they represent the physical bodily remains of our ancient ancestors.

2. Human mummies have been used to discover much information on the diseases suffered in antiquity which has then been applied to modern-day societies and their health problems.

3. Human mummies were very popular souvenirs collected in Egypt by wealthy travellers; therefore there are a surprising number of them located around the world. They are often associated with important information which helps to present a context in which they can be studied – for example inscriptions, artwork, artefacts and possibly even the tomb in which they were buried. They can provide a complete picture of material which can enhance our knowledge about the civilisation in which they played a part.

4. Family relationships between human remains can sometimes be deduced, such as in the case of the Royal Mummies, enabling the remains to be investigated as a small part of a larger population.

The fact that a great deal of research has been done on the human remains has led to a wealth of literature about them, whereas conversely, relatively little has been written about the animal remains other than excavation reports and knowledge of the fauna itself. This means that when one attempts to find information about animals in ancient Egypt, it is necessary to look at what has been done on the human remains and the techniques that have been used to study them. By looking at the human studies, it is possible to visualise what can be done by way of research on the animals.

The analysis of human remains has essentially acted as experimental research which has successfully identified which techniques have potential in this area before the attention can be turned to animal mummies. This is not to say that research on human remains has been exhausted as this is clearly not the case and interesting projects are constantly being undertaken, furthering our knowledge of the civilisation in which they lived. However, the majority of biomedical techniques in existence today have been attempted on human remains and it is difficult to develop a technical research project that represents totally new ground. The opposite is true for the animal remains as very little has been done, making them something of an unknown quantity and leaving research very much open as to what can be done and the results that can be achieved. It is with this in mind that I decided to study mummified animal remains (Owen 2001, 15).

The literature that does exist on mummified animals tends to belong to one of two groups – reports based on archaeological excavations and autopsy reports.

2.1 – THE EXCAVATION REPORTS

A number of archaeological excavation reports and interim reports in journals, some of which are described below, focus on the animal cemeteries where cult animals and votive offerings were buried. These provide information on the archaeological contexts in which mummified remains have been discovered although they do not provide much information on the mummies themselves beyond how they were buried and the containers they were buried in. They are nonetheless a good source of contextual background information which it is important to consider before progressing to the biomedical analysis of the remains themselves.

The Sacred Animal Necropolis at Saqqara

The necropolis at Saqqara was perhaps the largest animal cemetery in ancient Egypt and contains the burials of millions of animals, mainly ibis, falcons and baboons.

Saqqara is also the location of the Serapeum, the burial place of the Apis Bulls, which has been extensively excavated to reveal important information about the cult animals and the treatment they received after death.

Archaeological excavations at Saqqara were carried out by the Egypt Exploration Society during the mid twentieth century under the supervision of the eminent archaeologist W.B. Emery who published frequent interim reports detailing his findings in the Journal of Egyptian Archaeology. The excavations found evidence for the Apis bull burials alongside those of ibis and baboons, and they also located the resting place of the Isis cows, mothers of the sacred Apis bulls. The catacombs were found to consist of long corridors and galleries cut out of the rock, then lined with fine quality limestone blocks (Emery 1970, 7).

The 1968 excavation season focused on the burial chambers in the Serapeum which had been discovered in 1716 by Paul Lucas. Lucas had merely described what he had seen there without carrying out any archaeological excavation. Emery confirms Lucas' observations and locates the embalming workshop where the Apis bulls were mummified, including several "dummy mummies" constructed from wood and gesso, painted to resemble the features of the Apis (Emery 1968, 33).

The baboons appear to have been mummified by wrapping in lengths of linen fabric before being placed in wooden chests which were then filled with a substance similar to gypsum plaster or cement. The wooden chests were then placed in niches cut into the walls of the corridors and galleries before being sealed with a slab of stone. Interestingly, Emery reports on the presence of demotic inscriptions on these slabs which included the date of burial and a prayer for the "eternal welfare" of the animal, with some also giving the name, and where the animal was purchased (Emery 1970, 7). It is a rare treat for archaeologists to find such details on the animal recorded alongside the burial, which makes the discovery of the Saqqara complex so important. This is something which sets the cult animal burials apart from other animal mummies – largely because there was only one animal at any one time and each was viewed as having a 'personality'.

Emery believes that the baboon catacombs at Saqqara were originally constructed to house human burials possibly as early as the third dynasty – a theory he developed after locating a stone sarcophagus which inscriptions revealed to have been the resting place of a person who died around this time. This burial had been desecrated and the chamber later used to house baboon burials. Interestingly, the Upper Baboon Gallery at Saqqara was constructed under the mastaba tomb of a nobleman by the name of Kha-bau-Sokar of the third dynasty, which involved tunnelling deep into the rock beneath an existing tomb (Emery 1970, 7-10). The excavations of the ibis catacombs showed that they had been constructed on a site which already contained numerous archaic human burials. Emery speculated that the area must have had some significance associated with the ibis itself, otherwise it seemed like the builders had made life unnecessarily difficult for themselves (Emery 1965, 8).

The animal catacombs at Saqqara were elaborate subterranean galleries built on a colossal scale – those built to house the falcon mummies measured three metres high by two and a half metres wide, cut from solid rock. It is thought that there were many thousands of falcons interred in these chambers in sealed ceramic pots which had been stacked up against the walls from floor to ceiling. An inscription stela reveals that the catacomb was in use during the "Year 26, which makes Hathor Year 29 of Pharoah, by the servant Nekhtymneve" which corresponds to the year 89 BC (Emery 1971, 4-6).

Excavations at the ibis cemetery at Abydos found hundreds of birds interred in large clay pots dating from the Roman period. Associated with the pots were found brick enclosures holding the remains of other animals such as oxen, sheep and dogs, illustrating that although the cemetery was primarily for the ibis, other species could be buried there too. Peet states that,

> "The majority of the mummies were of the ibis itself or of bundles of its feathers, which seem to have been collected as sacred like the bird itself and mummified. Other mummies were those of various kinds of shrews, snakes and even of scarab beetles. Three small pottery jars were found to contain masses of broken ibis eggs."
>
> (Peet 1914, 39)

Archaeological excavations and scientific analysis of the material recovered at Saqqara are ongoing and benefit from funding from the Egypt Exploration Society. This work is continuing to provide answers to questions raised about what is surely one of ancient Egypt's most well-known burial sites. Human burials at the site date from the Early Dynastic and Old Kingdom, through the Amarna Period and beyond, with many of the royal family and their officials being laid to rest at the site. Saqqara is the location of the first pyramid built by King Djoser during the Third Dynasty (2649 – 2575 BC) and is also believed to be the burial site of the pyramid's architect, the Vizier Imhotep. The extensive network of animal catacombs and the areas dedicated to the Apis Bulls continue to fascinate scholars interested in ancient Egyptian animal cults (Nicholson 2005, 44).

The Sacred Animal Necropolis at Tuna-el-Gebel

Egyptologists believe that the establishment of a burial ground for the sacred ibis took place around 450 BC close to the ancient town of Hermopolis Magna at the site of Tuna-el-Gebel. It was customary practice for all dead ibises, irrespective of how they had lived or died, to be brought to the cemetery for burial by servants. Many of these early inhabitants of the cemetery had accompanying inscriptions which state where the ibis came from, the date it arrived and the name of the person who presented it for interment. These animals would have been mummified in their town of origin for practical reasons and those considered to be high status were placed in recesses in the walls of the catacombs which were then sealed with mud bricks. Archaeological evidence suggests that this practice continued up until the late Ptolemaic period; however the remains also suggest that the standards of care taken with the ibis at the cemetery declined steadily. Early burials show that each ibis received an individual burial, whereas in the later years, multiple birds were interred together, many without the dignity of any formal inscription (Kessler and Nureddin 2005, 125).

It is thought that a sacred colony of ibis were kept in the locality of the temple and it was from this group that special birds were chosen to fulfil ritual roles during temple ceremonies. These animals were involved in parades and specially staged appearances at the temple where visitors were able to witness the subjects to which the temple was dedicated. A limited number of animals that displayed certain coloured feathers and specific attributes were ritually slaughtered as offerings to the deities (Kessler and Nureddin 2005, 126).

The worship of the sacred ibis appears to have changed during the Ptolemaic Period when it was no longer customary for all the mummified birds to be transported to a major centre of worship such as Tuna-el-Gebel. Instead the birds were interred locally at a number of animal cemeteries. There have been suggestions that there could have been at least twelve or more ibis breeding grounds in the vicinity of Tuna-el-Gebel alone. Kessler and Nureddin (2005, 142) believe that this change was in connection with a new style of mummification with each bird being embalmed separately rather than being in response to the marked increase in the popularity of purchasing a mummified animal for personal devotion.

Tuna-el-Gebel served not only as the final resting place for the sacred ibis. In total, some ninety species of wild mammal, fish, bird and reptile were interred at the cemetery. Kessler and Nureddin (2005, 155) differentiate these specimens into two categories. Firstly are the animals carefully selected to take part in rituals at the temple complex, many of whom were slaughtered. Despite the often cruel methods of dispatch practiced, the physical remains of the animal were always deified by being preserved and taken for ritual interment at the cemetery. The second group of sacred animals were raised, lived and died at the feeding enclosures located near the temple, but did not display the special appearance required to ensure a role in the rituals themselves. These animals too were buried as if they were gods themselves.

Kessler and Nureddin (2005, 156) speculate that some fifteen thousand ibis mummies were transported to Tuna-el-Gebel annually from the local breeding grounds. They discuss the presence of numerous mummies mainly dating from the Ptolemaic Period which were found to contain either parts of an animal or associated items such as feathers, isolated bones and even nests. They consider that these mummies represent diligence on behalf of the embalmers in collecting every piece of material which was connected to the lives and deaths of the sacred birds, rather than being an early example of ancient fraud. Evidence suggests that when the birds were treated at their town of origin following death, the bodies would be left to skeletonise in large ceramic jars before the bones could be collected and wrapped together for transportation to the main cemetery. The disarticulated bones would be mixed together with those belonging to other animals before wrapping which explains why such varied assemblages of material have been discovered.

The mummies dating from Persian times found at the cemetery indicate that turpentine was used to increase the speed of the maceration of the flesh. This would have been applied in solution to the exterior of the body; however the use of this substance leads to poor quality mummies as it essentially burns the corpse leaving only a dust-like residue (Kessler and Nureddin 2005, 157).

During the Saite period, it seems to have become common practice for the eggs of the sacred ibis and other birds found at the breeding enclosures, to be transported to Tuna-el-Gebel for interment, often mixed in with actual mummies. It is unknown as to whether the eggs that were collected were fertilized or not; however it seems that they too possessed some special ritual significance in the minds of the ancient Egyptians (Kessler and Nureddin 2005, 159).

The baboon burials at Tuna-el-Gebel are interesting in that a number of pathological cases have been identified from the physical remains. The majority of the baboons died at young ages indicating the conditions of their captivity were not ideal. Many show evidence of conditions connected to limited space and insufficient nutrition such as rachitis and tuberculosis. The poor condition of the bones themselves indicates that the animals lived stressed lives which affected the thyroid gland (Kessler and Nureddin 2005, 160).

The Tomb of Sen-mut at Thebes

The tomb of Sen-mut at Thebes was found on excavation to contain the burial of a horse in its own coffin alongside the human burial. Horse burials in Egypt are relatively rare as horses were not indigenous to the country and their presence was limited to the upper echelons of society who had links to trade systems with other countries. This particular animal of the Arab type, which has been aged at between five and six years at death, was buried alongside his master, still wearing his saddle-cloth indicating that he was ridden during life. The burial is thought to date back to the early fifteenth century BC. Another horse burial (a ten-year old male) has been discovered at the eighteenth dynasty necropolis at Soleb (Clutton-Brock 1974, 93-94; Phillips 1948, 14-15).

2.2 – AUTOPSY REPORTS

In the same way that human mummies were unwrapped, so too were animal ones, although admittedly not amidst the same media frenzy. It can be presumed that over the years countless numbers of mummified animals have been studied in this way, mainly prior to the advent of biomedical techniques which no longer require the destruction of the mummy. A number of autopsy reports record the findings of such studies, some of which are listed below.

Brier and Bennett describe the autopsy that they conducted on an unprovenanced mummified fish purchased in Egypt to establish the species and the method of mummification employed. The remains were in a poor condition after having been eaten away by insect larvae (*Anthrenus* and *Attagenus*) which feast on desiccated animal matter, therefore making it impossible to establish whether the fish had been eviscerated. The examination did, however, locate a cut along the belly which may have represented an evisceration incision, into which mud had been packed to keep the shape. This fish, identified as a Nile catfish (*Bagrus bayad*), had been wrapped in a single piece of cloth, with bitumen soaked strips being used to secure it in place (Brier and Bennett 1979, 128-129).

Leek (1976, 131-133) describes the autopsy he conducted on a schilboid Nile catfish (*Eutropius niloticus*) following analysis using radiological techniques. He questioned whether the mummy was actually of a modern catfish as few other examples of this species have been located, but concluded that the method of mummification used was so similar to that displayed at the fish cemetery at Esna, that he believed it to be an ancient one. Analysis revealed that the fish had been placed in a layer of vegetable fibres belonging to the genus *Cyperus papyrus L.* before being wrapped.

2.3 – THE HISTORY OF MUMMY STUDIES

Before considering how modern multi-disciplinary approaches have changed the course of Egyptology, it is important to examine the type of research which was done prior to this in order to gain an understanding of the subject's history. In the early years, research on mummified material mainly revolved around the historical context in which it was found, to the point that only minimal scientific investigation was conducted.

The history of the study of mummified remains is closely related to the history of anatomy – the developments which affected anatomical research ultimately led to the rise in interest and the formulation of techniques for the study of mummies. The Renaissance saw the revival of interest in human anatomy and a move towards human dissections and autopsies which aimed to discover the intimate systems and workings of the body. This period was the first in which routine post-mortem examinations were carried out, mainly at the major Italian universities of Padua and Bologna. Prominent personalities of the period such as William Harvey and Michelangelo Buonarroti learned much about the functions of the human body by watching and conducting both human and animal autopsies. The Renaissance followed turbulent periods in world history during which only limited discoveries had been made as research on human remains had often been outlawed by the rulers of the time (Aufderheide 2003, 5-6).

The eighteenth century witnessed a massive surge in popularity in the study of pathology – the study of disease. Autopsies were carried out on the recently deceased, but it is known that some such studies were conducted on mummified remains, many of which were from ancient Egypt. An article appeared in the Philosophical Transactions of the Royal Society of London in 1794 entitled 'Observations on some Egyptian mummies opened in London' by Blumenbach (Aufderheide 2003, 7).

The nineteenth century saw some major developments in the study of mummies. One of the major protagonists of the study of ancient Egypt was Napoleon Bonaparte who enlisted a troop of one hundred scientists and historians to travel to Egypt and record everything they possibly could about the ancient civilisation. The work of these specialists culminated in the publication of nineteen volumes entitled 'Description de l'Égypte' (1809). This work, combined with the decipherment of the hieroglyphic code by Champollion and Young using the Rosetta Stone, increased the popularity of the study of ancient Egypt (Aufderheide 2003, 8-9).

During the nineteenth century much of the research into ancient Egypt revolved around the unwrapping and autopsy of human mummies. Invitations were sent out to distinguished members of society, inviting them to witness the unwrapping of the physical remains of an ancient individual. These 'unrollings' appealed to the upper classes of society as they were educational events where people went to see and be seen. They helped to increase awareness of the study of ancient Egypt and its people, although ultimately these events led to the destruction of a huge amount of irreplaceable material. However, not all of these unwrappings were merely social events. Dr. Thomas Pettigrew, professor of anatomy at Charing Cross Hospital, London, and John Davidson were the first to establish an educational course in Egyptology (1833) where students learned about mummification and how it was performed. The course involved unwrappings and autopsies, but they were performed by a doctor who specialised in anatomy, not simply an interested individual, so they can be seen to represent a major step forward in research (Aufderheide 2003, 9-10).

The twentieth century witnessed the turning point for Egyptology – autopsies gave way to biomedical study methods as the necessary scientific improvements took place. As new techniques emerged, Egyptian mummified remains provided an existing body of material which required study and which did not involve experimentation of these new methods on living patients. Today, we know the incredible value of these remains, yet one hundred years ago, the emphasis was on advancing science and mummies which had been brought over from Egypt by travellers as souvenirs seemed like the perfect material on which to experiment. The work on mummies from the twentieth century had success in two ways – it tested the new technological advancements and it raised interest in mummified remains.

Egyptology in Manchester

Manchester has a long and fruitful history at the forefront of modern Egyptological studies with the Manchester Museum housing a world-renowned collection of Egyptian material. A major patron of the Manchester Museum, Jesse Haworth (1835-1921), offered considerable financial support to British archaeologists working in Egypt during the late nineteenth and early twentieth centuries. Haworth, a prosperous Manchester textile manufacturer, travelled to Egypt in 1880 and as a result became increasingly interested in the study of the ancient Egyptian civilisation. He was responsible for funding Sir William Flinders Petrie's excavations in return receiving a share of the excavated material. Haworth then generously donated this material to the Manchester Museum and was responsible for funding an extension to the Museum built in 1912 which bears his name (http://museum.man.ac.uk/collections/egyptology/egyptology.htm). In total, the Egyptology collection contains in the region of 20,000 individual objects ranging from pot sherds to mummies. The Museum also received material from the excavations of the British School of Archaeology in Egypt, the Egypt Exploration Society, and the Liverpool School of Archaeology.

Manchester has become a centre for outstanding research following the pioneering work done by the first curator of Egyptology at the Manchester Museum, Dr. Margaret Murray, in the early years of the twentieth century. Murray was the first person to study mummified remains with the help of specialists in scientific fields. Prior to this major step forward, Egyptologists had tended to follow a purely archaeological or historical approach to the material, not wishing to be 'baffled' by the science which to them was often so alien to their way of thinking, that it was considered inadequate to explain the complex theories which they advocated. In Murray's time at the Museum, much of the research into the mummies still involved them being unwrapped and essentially destroyed, with the 'science' coming later in the investigations (David 1979; David 1986, 76; Owen 2001, 12).

Rosalie David arrived at the Manchester Museum in 1972 and began her renowned career as head of the Mummy Project, a multi-disciplinary team of individuals, all of whom were specialists in their chosen fields. The Project was established with two main research aims – 1) to identify evidence of disease, causes of death, and information on the living conditions and funerary beliefs and, 2) to establish a methodology for use on mummified remains which could be used around the world (David 1979, vii; David 1986, 76).

As the scientific methods evolved and technological advancements were made, it became less of a necessity to unwrap the mummies in order to study them. Egyptology had taken a massive step – to show faith in the new biomedical techniques which were revolutionising science, and to incorporate them into the study of ancient Egyptian remains. These techniques, when used alongside the historical, documentary and artistic evidence, could provide a complete picture of how the ancient Egyptians lived their lives and how they viewed death.

Egyptology at the University of Manchester received an enormous boost when the KNH Centre for Biomedical Egyptology was opened in 2003. It has become a pioneering academic department with a unique approach committed to the study of ancient Egypt through the use of multi-disciplinary approaches and the most advanced scientific techniques.

Radiology Case Studies

As I have previously mentioned, radiology has been applied to the study of ancient Egyptian mummified material almost as soon as it was developed, and there have been countless studies which have used these techniques to discover information about this type of material. As one would expect, the earliest of these studies were quite basic in terms of the results they managed to achieve simply because the technique itself was not advanced enough to enable anything more. Nonetheless, these studies represent the first non-destructive investigations to be carried out on Egyptian material and the initial step away from performing autopsies as a matter of course. Many of these studies have focused on mummified human remains and the ways in which radiological methods have been applied to them, but I am going to attempt to concentrate on those that have looked at animal remains. Where certain techniques have not yet been tried on animals, examples using human material will be incorporated.

Radiographic Analysis of the Feline Mummies in the British Museum

Armitage and Clutton-Brock (1981) conducted a landmark study on the remains of fifty-three cats dating from between 664 BC and 343 BC held at the British Museum, London. They used conventional radiology (hereafter referred to as X-ray) to determine that all the specimens had been embalmed in exactly the same way – eviscerated, the body cavity packed with sand and mud, then wrapped in linen. Electron microscopy was used to test tissue samples taken from the cats which revealed that natron had not been used directly on the corpse, although infra-red spectroscopy of the hair showed natron crystals which was interpreted as the bandages having been soaked in a natron solution prior to the animal being wrapped. Further analysis of the hair with comparative collections identified the species as being *Felis silvestris libyca.*

Armitage and Clutton-Brock were able to ascertain ages at death for all fifty-three cats using dentition and epiphyseal fusion to highlight two peaks at which most of the cats had lost their lives. Twenty of the cats were aged between one and four months, and seventeen were aged between nine and twelve months at the time of death. This suggests that the cats were deliberately killed at distinct ages, a theory that is reinforced by the fact that all except one were completely healthy. This individual showed signs of juvenile osteoporosis probably caused by a diet high in meat (No. 79.5351). Seven of the cats showed signs of vertebral displacement which could point towards strangulation as cause of death (Armitage and Clutton-Brock 1981, 185).

Previous Research by the Author

Previous research on animal mummification carried out by the author has largely revolved around radiological analysis on collections from both Bolton Museum and the Manchester Museum. These research projects were conducted whilst studying at York University and Manchester University and they represent the basis for this project. Information on these projects is included here as they demonstrate the type of methodology which has been formulated to date (further information can be seen in Owen 2000 and Owen 2001).

My first project radiologically examined thirty-nine previously unstudied animal mummies from Bolton Museum to produce a comprehensive catalogue of the material. The results showed an interesting phenomenon which has been demonstrated in previous studies – that of 'pseudo-mummies'. These mummies are wrapped to look like an animal, but on examination reveal only part of an animal inside, or in some cases absolutely nothing at all. In fact in the case of the Bolton mummies, a total of one third of the total assemblage were forgeries (Owen 2000; Owen 2001).

The Manchester Museum mummies also revealed upon radiological examination a number of pseudo-mummies, adding more weight to the argument that in many cases the embalmers were rather liberal with the truth, to the extent that the pilgrims were being sold fraudulent goods against their knowledge. It is also possible that the ancient Egyptians considered any part of an animal to be just as special as a whole animal which would justify this practice. This could be demonstrated by the findings at Saqqara of ibis eggs and feathers mummified alongside the complete animals (Emery 1970). I personally find these pseudo-mummies very interesting as they provide an interesting insight into the votive mummy industry (Owen 2001, 104-106).

A Feline Mummy from Niagara Falls Museum, Ontario

A cat mummy from the Niagara Falls Museum, Ontario, was studied by Pahl (1986, 133-140) during the 1980s using X-ray which enabled an identification of species to be suggested for the skull. The post-cranial skeleton appeared to be missing, but interestingly human bones had been added to provide the shape of the cat before wrapping. These included a length of tibia with corresponding fibula which had been inserted into the foramen magnum of the feline skull. It appears that the embalmers were not opposed to including human body parts inside what was intended to be an animal mummy.

The Animal Mummy Project, Cairo Museum

Apart from X-ray, minimal work has been done to date using radiological methods on mummified animals. This is likely to be due to the fact that human mummies are generally the initial recipients of advancements in biomedical techniques. Numerous studies have been done on the human remains using computed tomography (CT) which has provided fantastic results, but as yet animals have not been studied in this way, with the exception of a couple of mummies being studied in Egypt by Salima Ikram at the American Research Centre in Cairo (Ikram 2005, 207-227). The simple fact is that nobody has, until now, viewed the animal remains with enough interest to attempt these costly procedures in any detail, and using the very latest radiological equipment. Hopefully this research project will elevate the study of these animal remains to the same standard as their human counterparts through the use of CT.

The Animal Mummy Project was established in 1998 by researchers Salima Ikram and Nasry Iskander who aimed to study the animal mummies held in the Cairo Museum collection. An ultimate aim of the research was to stabilise and conserve the mummies, which can be difficult if the items are on permanent display to the general public. Fortunately, the project coincided with the transportation of the mummies from their original display gallery into museum storage, and their subsequent installation in the specially renovated Animal Mummy Room at the Cairo Museum. The mummies were studied using visual examination and X-ray alongside information gathered from surviving documentation. CT was not used during the project due to a lack of facilities (Ikram 2004a, 4).

Ikram and Iskander have succeeded in publishing two popular catalogues of the animal mummies held at the Cairo Museum which are currently on sale to the visiting public. These provide a general introduction on the mummification of animals followed by a brief catalogue of the mummies, accompanied by photographs and radiographic images.

The Animal Mummy Project has achieved what it originally intended to do. The animal mummies have been studied before being conserved, stablilised, and placed on display in their new exhibition room. The information gathered from the study of the mummies revealed a number of interesting points.

In her report, Ikram discusses the evisceration of animals during the mummification process, drawing parallels between them and the human mummies. With the human mummies, it was usual to retain the eviscerated organs, either by wrapping them and placing them in canopic jars or by treating them and returning them to the body cavity. With animal mummies, whether the corpse was eviscerated depends on the type of mummy being made. The cult animals were generally eviscerated and the organs were stored in canopic jars or placed in the body cavity, similar to human mummification. Certain victual mummies, especially poultry, appear to have been eviscerated as well, but generally the organs were discarded (Ikram 2004a, 14). The votive mummies are generally not eviscerated at all. This was probably because they were manufactured hastily to meet the enormous demand, and evisceration was an unnecessary procedure bearing in mind the poor quality of the mummification technique as a whole.

Ikram mentions the frequent occurrence of false mummies which are found to contain no animal material. She believes that the ancient Egyptians may have considered that these specimens were still sacred due to their association with a place of worship (2004a, 17). This is certainly a possible explanation for the high volume of these pseudo-mummies although the lack of available documentation means that they will never be completely explained.

For those mummies in the Cairo Museum collection which were found to contain skeletal material, Ikram makes some interesting observations. The first specimen is that of a pet monkey (CG29837) which was discovered in a tomb believed to belong to Amenhotep II who reigned in Egypt from 1427 – 1401 BC. The radiographic investigation of the monkey revealed that the animal had died before it had reached the age of two and a half due to the stage of epiphyseal fusion. Ikram notes that the internal organs are not visible on the radiographs which, she interprets, as evidence for evisceration per anum (the organs having been dissolved in situ through the injection of solutions into the anus) (2004a, 22). This statement could be challenged by stating that it is unlikely that any organs would be visible on the images due to the effects of autolysis on the soft-tissue. It is undoubtedly the case that beloved pets, mummified to ensure they achieved an eternal life, were eviscerated in many cases. The apparent absence of internal organs on the radiographs undoubtedly merits further investigation. The application of CT scanning to the monkey would rule out the possibility that the organs could have been left inside the body cavity, but that the desiccation process has made them difficult to detect. CT would enable the investigator to view axial sections through the monkey which would either prove or refute the presence of internal organs.

Ikram also notes the case of a votive crocodile mummy (CG29712) which had been crudely mummified through the application of resin. The radiograph appears to show that some of the bones within the corpse have been destroyed and she believes that this may represent combustion caused by the hot resin.

> 'It is possible that the extreme heat of the resins applied to the corpse of a young crocodile, combined with the natron and other materials used in mummification, caused the animal to combust and the bones to disintegrate, or, perhaps it was made when there was a dearth of crocodiles'.
>
> (Ikram 2004a, 17)

It seems far more plausible to believe that the votive crocodile mummy in question never actually contained any skeletal material. Numerous false crocodile mummies have been discovered in animal necropolises across Egypt so it seems realistic to presume that this specimen is simply another of the same.

Animal mummies are important in highlighting the demographic fluctuations of the breeds themselves. It is important to remember that there was a far greater range of animals in ancient Egypt than there are today. The changes which have occurred have been due to climatic and environmental alterations and the impact of man. A number of breeds which were common in ancient Egypt have migrated farther south as the climate has changed and, as the human population has increased, the land available to them has proportionately decreased, forcing them out into other areas (Ikram 2004b, 4).

Ikram discusses the appearance of victual animal mummies – those placed within the tomb as food offerings for the deceased. To give scale to this practice, Tutankhamun is reputed to have had close to forty boxes of victual offerings placed within the confines of his tomb chamber. The most commonly identified victual animals appear to have been waterfowl (namely geese and ducks), pigeons and cattle. The Cairo Museum collection contains a number of animal mummies which are believed to have been intended as victual offerings. A number of these are disarticulated joints of meat, ribs (CG51098) and legs (CG29853) which have a very different appearance to the animal mummies which are intended to represent whole animals. The victual mummies are quite obviously disarticulated joints of meat, wrapped simply in linen bandages and presented on offering platters. There are not manufactured to resemble an animal so they are easy to differentiate (Ikram 2004b, 16-17).

Ikram and Iskander's research at the Cairo Museum has succeeded in raising the profile of animal mummies and the information that can be gleaned from them through the application of X-ray. It has, however, got a number of flaws which should be addressed here. Ikram's publications include a number of observations which are based on little more than assumptions. It is important to stress that a number of these comments could be relatively easily answered with the use of further radiographical analysis such as CT. I think that the publications made on behalf of the Supreme Council of Antiquities are intended to make the study of animal mummies popular with the general public. This in itself is important as they tend to get less attention than the human mummies, so it is necessary to raise awareness of their existence and highlight the information that can be gathered through their study.

The Animal Mummies in the National Museum of Antiquities, Leiden

Falke (1997) conducted a radiographical study of twelve animal mummies from the National Museum of Antiquities in Leiden, the Netherlands. The mummies studied (of a variety of species) were all thought to be Graeco-Roman in date; however, no information was known about where they were found. Falke's study is perhaps the most comprehensive example discussed in this work as it incorporated as it incorporated X-ray, CT, multiple beam equalization and magnetic resonance imaging (MRI).

The collection studied by Falke included four mummies all of which were thought to be cats. Radiology confirmed that three were well preserved complete cats, but the fourth mummy was found to contain only the skull, proximal section of the vertebral column and the forelegs of a cat. Falke reports that the earliest examples of mutilated or pseudo-cat mummies were described around 1930, a number of which were located in the collections of the Museo Arquelogico, Madrid (Falke 1997, 56).

Interestingly, the cat mummies studied, all belonging to the sub-species *Felis lybica*, displayed potential pathological anomalies. The first note that Falke makes is that all four cats display evidence for thin cortical bone when compared to modern feline radiographs. This could be an effect of time and the mummification process on the mineralisation of the bones, or it may be a direct correlation with the nutritional well-being of the animals in life. The bones of one cat showed considerable under-mineralisation which was interpreted as the result of a diet high in offal fed to the animals in captivity which caused high phosphorus levels in the blood and consequently a calcium deficiency. The nutritional quality of the diet was also given as the contributory factor in causing the fracture of a femur in one of the cats (Falke 1997, 56).

The age at death was calculated for the cats using the rate of dental eruption and epiphyseal fusion which indicated that all four cats were immature. Falke mentions the study carried out by Clutton-Brock and Armitage at the British Museum, London, which isolated two age ranges in which cats seem to have died: one – four months and nine – twelve months (Armitage and Clutton-Brock 1981). These ages are displayed through radiological evidence and indicate that the ancient Egyptians must have practiced deliberate slaughter to ensure that there was a constant supply of animals to mummify for votive purposes. Falke (1997, 56) states that no cause of death could be ascertained using the radiographs of the Leiden cats so it was not possible to suggest how they were dispatched. There was no evidence for neck fractures as identified in the British Museum study.

The Bubasteion, Saqqara

Research carried out on the animal mummies excavated from the Bubasteion, a cat necropolis dedicated to the goddess Bastet situated at Saqqara, between the years of 1991 and 1999 analysed 272 so-called 'packets'. Zivie and Lichtenberg opted to call the specimens by this name as it was felt to be more appropriate for the material judging from the high volume of pseudo-mummies encountered. They comment that the 'fake mummies' were easily recognisable as they tended to measure around twenty centimetres in length and were much smaller than the true mummies (Zivie and Lichtenberg 2005, 114).

A mobile X-ray machine was used to image the animals on site, with the smaller specimens being studied in batches of up to six at a time. Those measuring over sixty centimetres in length had to be radiographed in sections with an overlap and joined together at a later date (Zivie and Lichtenberg 2005, 115). A detailed photographic record was also produced at the same time for each mummy bundle. CT was not attempted.

The Bubasteion study revealed some interesting results worthy of mention in this project. In total, sixty-eight percent of the assemblage studied was found to contain feline skeletal remains, four percent yielded isolated feline bones only and thirty-one percent were found to be pseudo-mummies (Zivie and Lichtenberg 2005, 115).

Zivie and Lichtenberg identified two different body positions given to the mummified cats from Saqqara. The first they term 'skittle mummies' due to their cylindrical appearance, with the second type displaying all four limbs wrapped individually as if the animal was lying on its side. It is noted that although mummies displayed in this position are found in the collection at the Louvre Museum, Paris, none of the cats studied by Clutton-Brock and Armitage from the British Museum are in this pose. At the site of Saqqara, the animals studied appear to show that both methods were widely used with no one method being a clear favourite (Zivie and Lichtenberg 2005, 116).

The mummies studied have been simply desiccated with no evidence of evisceration. The so-called 'skittle mummies' are generally in a better state of preservation possibly due to their more robust shape. None showed signs of elaborate decoration although a small number of mummies were presented in sarcophagi made from stone or wood. The hypothesis is made that these cats may have been considered to be the manifestation of Bastet herself and therefore deserving of a more elaborate burial (Zivie and Lichtenberg 2005, 116).

The study revealed extensive evidence to suggest violent fractures of the cranial bones, in particular those in the occipital region, and dislocation of the cervical vertebrae suggestive of death by strangulation. In total, two thirds of the collection showed signs of having suffered a violent death and only one third appeared to have survived to reach adulthood (Zivie and Lichtenberg 2005, 117).

Commenting on the existence of the pseudo-mummies within the collection, Zivie and Lichtenberg (2005, 118) say:-

> 'the number of 'fake' offerings would suggest an excessive demand and an insufficient supply of cats, rather than a desire for deliberate trickery on the part of the priests'

They make the point that those mummies which contain skeletal remains of a far smaller proportion to the bundle itself may represent a desire to mislead the pilgrims.

Computed Tomography (CT) of Human Mummies

Discussed below are a number of research projects that have utilised the method of CT to study human mummified material. They have been included here to give the reader some understanding of the vast amount of knowledge that can be gleaned from this material.

D'Auria and Marx conducted a research project between 1983 and 1987 using X-ray and CT to study a total of fifteen fully wrapped human mummies (1988). The CT data from one of the human mummies, Tabes, Songstress of Amun, was selected in the hope of formulating an experimental three-dimensional reconstruction of her body. The CT scanner produced two millimetre thin axial sections of the mummy at two millimetre intervals through the head and neck, and four millimetre sections at four millimetre intervals over the rest of the body. Research into the literature available on this subject suggests that this was the first investigation of its kind to use CT data to produce a three-dimensional reconstruction of a human mummy.

D'Auria and Marx's paper published in the American Journal of Radiology made the following point (1988, 147):-

> 'axial CT images provided a great deal of information of anatomic and Egyptologic interest, it did not reveal that which was of most importance to the ancient Egyptians, namely, how the mummy looked. It did not show how successful the art of the embalmer was and how his efforts had weathered the millennia'

Although I agree mostly with this statement, I feel that radiography can help to provide information on the work of the embalmer and the extent to which they achieved the goal of complete body preservation. CT images can be used to identify the mummification techniques used and the state to which preservation was achieved. D'Auria and Marx's statement was written when CT technology was being improved following its original introduction during the 1970s. This may account for the seeming lack of information that they report on the work of the embalmer.

The CT reconstruction produced excellent images of the surface of the mummy showing the subcutaneous packing materials which had been applied unevenly, leaving a bulge in the neck region. The orbital cavities are empty indicating that false eyes have not been applied in this instance (D'Auria and Marx 1988, 148). D'Auria and Marx's paper concluded with the following observations (1988, 149):-

> 'Three-dimensional images from axial CT data is a further advance in the non-destructive and non-invasive evaluation of Egyptian mummies, offering a different and perhaps more appropriate perspective of these objects. Surface details were apparent that were not appreciated on axial images. The use of medical imaging technology in non-traditional fields represents an opportunity for the exchange of ideas between historically isolated disciplines and allows the radiologist to make contributions to the understanding of our world and origins'.

In a study conducted by Hoffman *et al.* using nine Egyptian mummies from the Michael C. Carlos Museum, Atlanta, Georgia, CT scanning was used to establish whether the method had a valuable role in the study of mummified material. Seven adult and two child mummies were involved in the experiment: they were all imaged in one millimetre thick sections from skull to mid-cervical spine and three millimetre sections for the complete body. This technique produced around 1017 sectional images for each adult mummy and around 400 for each child. It was discovered that spiral CT of one to three millimetre section thickness gave a wonderful insight into the contents of the body cavities, which enabled the authors to differentiate between packing materials and internal organs. Three-dimensional reconstruction was also attempted, as was a virtual fly-through using techniques currently being used on patients undergoing colonoscopy and bronchoscopy examinations. The authors conclude by saying (2002, 385):-

> 'As multi-detector row CT and post-processing tools evolve, the capabilities and applications of CT will continue to proliferate, attesting to the expanded versatility and utility of CT as a non-invasive research tool in the multi-disciplinary study of Egyptian mummies.'

The radiological study of the Leiden collection also contained the mummified remains of a dog which is estimated to be adolescent in age as the last molar has not erupted. The mummy which was thought to be a gazelle was identified as being a pseudo-mummy and has since been compared to a similar mummy held at the Field Museum of Natural History in Chicago (Falke 1997, 57).

The worship and mummification of crocodiles began relatively late in Egyptian history with the animals being bred and raised in captivity at the temple complexes by the priests. Documentary records suggest that the priests had to exercise extreme caution when dealing with the crocodiles and that they employed special measures to ensure that their personal safety was maintained at all times. They appear to have restrained the animals during feeding time and essentially force-fed them a diet of meat, bread, honey and water (Falke 1997, 57). There was one mummified crocodile in the study collection; however the skeleton was incomplete following post-mortem mutilation. Falke discusses the frequent occurrence of pseudo-crocodile mummies including one example described in 1901 by de Goro Starzu which contained a crocodile skeleton with sections of a larger animal placed on top, with approximately fifty newborn crocodiles included in the wrappings. The crocodile mummified in this instance is likely to have been a mature creature to allow sufficient space to include so many immature crocodiles.

The two fish in the Leiden study collection display evidence that the animal had been sectioned longitudinally along the body to allow the mummification preservatives to penetrate the tissues and effectively dehydrate the corpse. The radiographs show a layer of a mud-like substance directly applied to the skin of the fish which would then have been buried in hot sand to aid the dehydration process (Falke 1997, 57).

Falke concludes his paper with the following statement which discusses the potential value and limitations of the methods used (1997, 58):-

> 'The advanced multiple beam equalization technique, by increasing the dynamic range of the film, offers the possibility of visualizing both the soft tissue and high density parts. This advantage is of particular interest when mummies with large amounts of overlapping opaque resin are present. CT further adds to evaluation by provision of true sections eliminating overprojection. This is especially valuable in the presence of radiographic embalming materials. CT is also helpful if multiple species are present. In our experience, imaging with MRI was not possible due to the absence of signal from dried tissues'

Experimental Mummification of Animals

Egyptologists have attempted to clarify how animal mummies were produced through the use of experimental archaeology where documented methods and evidence from the skeletal remains themselves are recreated. Attempts should be made to ensure that only materials and technology thought to have been available to the ancient Egyptians are used to ensure that the experiments are as historically accurate as possible. In the 1990s, Brier (1996) mummified a human body donated to medical research using evisceration and natron desiccation. The earliest known experimental mummification of animals was conducted by Lucas (1962, 290-295) when chickens were mummified in natron solution and common salt to assess the success of each as a dehydrating agent. Further studies were conducted at the University of Manchester (Garner 1979) and the American University in Cairo (Ikram 2005). The details and results of Ikram's studies are included below to provide the reader with an understanding of the difficulties encountered during the procedure of mummification.

Ikram's research in Cairo revolved around the experimental treatment of five rabbits, two ducks and a series of fish between 1999 and 2002 and was intended to test the mummification methods identified as having been used on animals. Attempts were made to source ingredients that would have been used in ancient times from herbalists and apothecaries in Cairo itself. The natron required for the experiments was ground using industrial equipment due to the large volume (in excess of seventy kilograms) that was required. Where possible authentic oils and resins were purchased from Ethiopia, Syria and the Sudan, which are all known to have been recognised trade routes during ancient Egyptian times (Ikram 2005, 30).

One rabbit was not eviscerated, but simply treated using natron and frankincense as a means of controlling the odour given off by the remains. Although the rabbit was placed in a controlled environment meant to replicate the humidity and heat of the desert, it had to be disposed of after only one day as it had essentially exploded through the actions of the internal gases and fluids present in the corpse (Ikram 2005, 31).

A second non-eviscerated rabbit was also treated by making a series of holes in the body intended to allow gases and fluids to leave the body and be absorbed by the natron. The corpse, packed in natron, was placed in direct sunlight on the roof, before being disposed of after seven days following excessive beetle infestation of the remains, the complete disintegration of the pelt away from the soft-tissue and the explosion of the head. Ikram believes that this animal would have eventually become naturally mummified had it been allowed to stay in direct sunlight, but 'health and sanitary considerations' forced the experiment to be concluded prior to this being achieved (Ikram 2005, 32).

The third rabbit to be studied was bled by severing the carotid artery and the internal organs were removed through a ventral incision with the exception of the heart which was left in situ. The remaining body cavity was cleansed using a local brand of gin as it was not possible to source palm wine which is thought to have been the substance used by the ancient Egyptians. It was then washed with water and dried as thoroughly as possible using linen fabric. The rabbit was balanced on wooden blocks in a container of natron with more of the substance being placed inside the body cavity. The container was again placed on a rooftop in direct sunlight and was checked every few days, with the natron being changed periodically when it appeared to be damp. Insects had begun to act upon the remains around the sixth day; however by day sixteen these were fairly limited in number. The rabbit was removed from the natron on day twenty as the remains were completely dry and bones had begun to protrude from the skin. Much of the fur had fallen off and was evident in the natron itself. The desiccated animal was then treated with castor oil with linen balls soaked in the substance being placed within the body cavity. Molten beeswax was applied over the ventral incision to seal it – a practice that has been witnessed in human mummies – before the rabbit was wrapped in linen and coated in molten resin (Ikram 2005, 33-34). Ikram notes how brittle the animal had become following the desiccation process and how extreme care had to be taken to ensure that it was not damaged in any way.

The fourth rabbit was eviscerated through an abdominal incision and the body was massaged to encourage fluids to leave. It too was washed with gin which was poured into the body cavity, swilled around and poured away. The body cavity was packed with natron before the animal was placed in a container and placed in direct sunlight. On day two, the natron inside the body cavity was removed as it had begun to smell and small linen bags full of the substance were placed in the cavity. Evidence of this practice has been witnessed at sites where embalmer's materials have been located and it appears to be a far more efficient method of using natron internally. The natron in and around the rabbit were changed regularly and the body was massaged to allow fluids and gases which had accumulated to escape. Ikram notes how areas of the body remained damp and started to become puffy over the course of the experiment, until the animal was totally dry on day twenty-seven when it was removed from the natron. Castor and lettuce oils were poured over the body to allow the joints to be manipulated and the animal placed in a more natural position. Molten resin was then applied to the inside of the body cavity and further bags of natron and chunks of resin were introduced to mimic those found in other animal mummies. Again the brittle remains were wrapped in linen and sealed with molten resin (Ikram 2005, 34-37).

The final rabbit was eviscerated by means of a turpentine enema rather than by an incision. Initially the corpse smelled relatively fresh in comparison to others, but by the seventh day the skin was detaching itself from the soft tissue so the linen bung was removed from the anus allowing the turpentine substance to exit. It was noted that the introduction of the fluid had destroyed a section of the hind leg, but that parts of the internal organs remained within the body cavity. On the ninth day, the body was massaged and chunks of the internal organs left the body. The rabbit was finally removed from the natron on day thirty-five and was again treated with castor and lettuce oils before being wrapped in linen with the addition of further rolls of fabric to support the neck area. Ikram noted that the state of the fur and the absence of any unpleasant odour in this case suggested that the desiccation process was more advanced; however the animal had taken on a compressed appearance as the majority of the viscera were no longer present (Ikram 2005, 37-38).

The fish mummies produced during the course of the experiment belong to the catfish (*Clarius*) and perch (*Lates niloticus*) families which have been commonly identified amongst ancient mummified specimens. The initial step was to gut the fish before tying the body with linen strips and placing in a container of natron. The natron was exchanged periodically as it became soaked with fluid from the corpse. The perch specimen was removed from the natron on day fifteen with the more greasy catfish being left a further two days. The corpses were treated with lettuce oil, wrapped in linen and secured with resin (Ikram 2005, 41).

The two ducks mummified during the experiment were eviscerated before being placed in containers of natron. The first specimen had fresh natron in the container every five days and the second duck was left in the same natron for the full length of the experiment. Both ducks were removed from the natron on day thirty-five with the one that had fresh natron every five days yielding more successful results. Both ducks have since been disposed of as the remains became odorous (Ikram 2005, 42).

Ikram's experimental work in Cairo has been successful in highlighting methods of mummification which achieved better preservation and also in demonstrating areas where problems occurred during the procedures themselves. The three rabbits which successfully survived the mummification process and both of the fish specimens remain in a stable condition today indicating that the techniques achieved desiccation to a sufficient state. It is hoped that in future years, the animals will be studied using radiographic techniques to assess the internal state of preservation. Samples may also be taken for analysis to test the state of degradation of the DNA and to evaluate the methods available for the extraction of DNA molecules from mummified animal material (Ikram 2005, 42).

2.4 – AIMS AND OBJECTIVES

In the study of ancient Egypt relatively little is known about mummified animals. As discussed previously, the primary focus of this project is the use of imaging to study the mummified animal material to provide answers for a number of proposed research questions. The limited amount of existing knowledge surrounding these specimens means that the subject could have been approached from a variety of academic directions. There are a number of biomedical research techniques which have shown potential in the study of human mummies and which could potentially be valuable applications to animal material in the future. My expertise and academic background has caused me to focus on certain aspects over others, but the opportunities for further multidisciplinary research are abundant and will hopefully provide many students in the future with exciting projects.

The following clearly defined research questions have been established for this project:-

1. Why did the ancient Egyptians mummify animals? – This relates to the important belief systems behind the

worship of animals in both life and death and can be seen to work alongside and in parallel with the practice of human mummification. This question encompasses many different aspects of the life of the ancient Egyptians and draws upon evidence from art, literature and the archaeological remains themselves.

2. How was animal mummification carried out? This question will provide different answers depending on the type of mummy being studied. For example, votive mummies display very different methods of mummification from those applied to cult animals and pets. This question will be answered through a close assessment of the literature concerning the autopsies of animal mummies, and through new evidence highlighted with the use of radiology.

3. Can CT be used successfully on mummified animal material or does it simply duplicate information gathered using X-ray? This question relates to the potential usefulness of the method itself and should be relatively easy to answer, following practical sessions.

For the purposes of this research project the decision was made to focus attention on the use of non-invasive techniques, namely radiology, for a number of reasons:-

1. Museums and other institutions that house archaeological collections containing mummified material are generally more inclined to loan artefacts for further study by non-invasive techniques. This is usually because the merits of any research proposal are generally considered by a management board who decide whether they are prepared to take part in the project and on what conditions material will be loaned. In my experience, non-invasive techniques will always be granted permission before invasive ones, often with the proviso that further analysis may be considered should the need for it be justified based on the results of the non-invasive techniques.

2. The use of non-invasive biomedical techniques on archaeological material is undoubtedly considerably more ethical than the use of invasive methods. This relates to the fact that all mummified material is a non-renewable resource and should be treated accordingly. Should the use of invasive methods be required in order to answer the particular research questions posed, sampling should be implemented on a minimal scale to reduce the impact of damage on the specimens. This is becoming easier as new technologies are introduced to minimise the sample size required in order to get satisfactory results.

3. The use of radiological techniques should constitute the initial phase in the analysis of mummified material as it allows a completely non-destructive first impression of the specimen to be gathered. In general, mummified animal material encountered in museums has never been studied on any level so it is important to begin with a thorough radiological assessment. This initial analysis may provide ample information with which to answer the research questions or it may highlight potential avenues for further research.

The research has a number of objectives which I hope will be addressed during the course of the project.

1. To assess the potential of radiological techniques in the study of mummified animal remains. Their potential is well-attested on human material, but will it prove to be as valuable with animal remains?

2. To produce catalogues of the mummified animal material for the loaning institutions so that they can update their records and, in many cases, display the artefacts alongside informative text based on scientific evidence.

3. To identify what mummified animal material is available in museum collections both nationally and internationally for the benefit of future researchers. At present, information on animal mummies held in collections is disparate and in the majority of cases, the specimens have never been studied beyond external visual observations.

This project focuses on the application of radiological imaging techniques to mummified animal material. In general, X-ray and CT were carried out on all the loaned material simply because it was felt that whilst the facilities were available, and the specimens were at hand, the opportunity should be seized. The information gathered from the two methods tends to complement each other rather than duplicate it, so it is worthwhile to carry out both methods if time and funding restrictions allow it.

The specimens were studied in batches in the Department of Clinical Radiology at Manchester Royal Infirmary. Four specimens were radiographed at York District Hospital. In both of these cases, the facilities have to be used outside of normal working hours (weekends or during the evening), unless emergency cases take precedence. In these scenarios it is necessary for the ancient specimens to make a hasty retreat and allow the living patients to be treated immediately. The National Health Service and independent hospitals are often very keen to get involved in studies of ancient material, but understandably they demand that living patients take priority.

The use of hospital equipment has considerable advantages for archaeological studies as it means that the very latest technology and the expertise of the medical staff are available. At the University of Manchester students are extremely fortunate to be linked with one of the countries leading teaching hospitals and also with the department of Imaging Science and Biomedical Engineering which is instrumental in pioneering the use of radiological techniques. This also applies to other biomedical techniques which can be applied to mummified material such as histology where specialists in these fields are willing and able to donate their time and expertise to the study of Egyptian material. For this reason, Manchester can be seen to stand alone in its multi-disciplinary approach to the study of ancient Egypt and is why the Manchester Mummy Project has become famous around the world. It is often enthusiasm and personal interest in this field which fuels these close relationships between academic departments enabling such pioneering research to be carried out.

Chapter Three
METHODS

3.0 – THE ARCHAEOLOGICAL RECORD

> 'By its very nature, the archaeological record is not a faithful mirror of the history of the site....... Consequently, our knowledge of animal use and the environments in which people lived is based on fragmentary, incomplete, and altered evidence.'
> (Reitz and Wing 1999, 326)

Archaeological material is inherently governed by a number of factors which have to be considered when conducting research. Archaeology deals with the material remains of past civilisations and the reasons behind the survival of certain materials over the passage of time can be affected by a number of factors. The material itself and location of its burial can determine whether or not an artefact survives. Undisturbed deposits tend to yield better results yet it depends on the type of soil or burial deposit as to which materials, if any, would survive. Ancient Egypt is extraordinary in that such a vast amount of the physical remains have survived in an excellent state of preservation and often for thousands of years. It can be tempting to view such a marvellous assemblage of material as being representative of all the original material, yet it must be remembered that the artefacts to which we have access are just a small fraction of what was originally there, and our material must be viewed in this way.

Although the climate of ancient Egypt has been paramount in ensuring the preservation of much of the material culture, there are situations where the material has been unobtainable or even destroyed by an opposing environmental phenomenon – flooding. The excavation seasons at the Bucheum in Armant, resting place of the cult Buchis bulls, have been continually plagued by flooding as the excavators Mond and Myers (1934, 57) describe:-

> 'Evil circumstances of unusual strength and number combined to prevent us from obtaining much valuable information about the mummies.'

Ancient Egypt has provided documentary and artistic evidence which can complement the archaeological record in helping to determine what the artefacts mean. This can be extremely important, but in many cases – especially those involving burials of humans and animals – there is little documented evidence to explain the remains, as the burials have often been subjected to plundering by grave-robbers and poor recording by nineteenth century excavators. There is very little that we can do to combat these inherent problems, as we have a duty to examine the material which remains, but it is important to reiterate that flaws do exist. As Shackley so delicately puts it, archaeologists are responsible for "finding some order in the chaotic archaeological record" (1985, 13).

Alongside acknowledging the fragmentary nature of the archaeological record, it is important to discuss the nature of the collections that Egyptologists have at their disposal. We have already addressed the fact that only a small proportion of the whole survived the passage of time to be preserved in the archaeological record. Egypt is fortunate to have a wealth of surviving material. However, the mere survival of the material is only the first step in establishing its potential for scientific study. Much of the material which survives will never be found, and of the small amount that is discovered, only a fraction of this will be excavated and will find its way into international museum collections and scientific studies.

For the purposes of this project, I have been extremely fortunate to find a number of institutions around the United Kingdom who have been supportive towards the aims of my research. At the outset of the project, concerns were raised as to the number of animal mummies that would be made available for study, especially given that the radiographic analysis would be carried out in Manchester which would necessitate the transportation of the collections over considerable distances in some cases. In total, I have been able to study four complete collections – Bolton Museum and Art Gallery, Manchester Museum, Bristol Museum and

Liverpool University. To complement these collections, I have also had access to isolated specimens and a small collection from Stockport Metropolitan Borough Council.

I was able to identify potential collections for study in this project were identified through direct mailings to all institutions in the United Kingdom which have Egyptian collections. Forms were sent for completion asking about the nature of the collections initially to ascertain the volume of animal mummies that are in the country. The museums that had animal material were asked to state whether any of the specimens had been studied in any way and whether they would be receptive to further study through this project. A minimal number of specimens had been studied, generally through radiography, with a couple having been subjected to invasive sampling for histological and DNA analysis. It should be noted that most of these studies were small-scale studies for undergraduate research projects. None had been published.

It is firmly believed that the vast majority of the specimens studied during this project belong to the votive animal mummy category. Many of the specimens studied do not have a great deal of supporting documentation to record their provenance and date of production so there are very few suggestions as to which category of mummy they belong. Those that do have provenance information suggest that they were removed from some of the largest and most well-known animal cemeteries in Egypt. This is further reinforced by the nature of the remains, the majority of which are relatively crudely mummified, suggesting low-cost production methods.

One of the major considerations which must to be taken into account by anyone studying mummified remains is that mummies are not a renewable resource. Any material to which we have access for study purposes is finite – once it has been destroyed, it cannot be replaced. This is a standard concern with all archaeological material, and it is the responsibility of Egyptologists to ensure that as much of this material as possible survives for the enjoyment and education of future generations. In many cases, the work of Egyptologists helps to preserve ancient materials by highlighting items that require conservation, and through the desire to protect the resources for the future. With this in mind, the multidisciplinary method of study has come to encapsulate a number of techniques which are absolutely non-invasive, or require only minor samples to be taken.

Egyptologists are responsible for studying, recording and preserving a finite archaeological resource. Unlike other scientific disciplines, it is not possible for mummified remains to be 'reproduced' in laboratory conditions, as it would be with experimental animals or ordered from a chemical catalogue for example, nor can experiments be designed to develop cures for the 'symptoms' of mummification. Egyptology remains today primarily as an exercise in observation. Artefacts are gathered from an ancient civilisation and studied using the scientific techniques available in the hope of increasing modern understanding of them. Established methods pioneered in other areas of science and medicine are required to be adapted for use on dehydrated and desiccated archaeological material. In some cases this requires significant remodelling and reworking of the existing techniques in order for them to be successful on mummified material, creating opportunities for inventive scientific experiments.

The study of mummified material can take on two different forms – invasive and non-invasive. Invasive research uses methods which rely on the removal of samples for analysis using scientific methods. The sophistication of these methods usually facilitates the extraction of minute samples (in many cases less than one gram in weight) but still the specimen is damaged in the process as it loses some of its physical integrity. Some scholars believe that the removal of any samples, no matter how small, from an archaeological specimen is destructive and for this reason does not constitute ethical research. Under certain circumstances, I believe that damaging specimens by taking samples can be justified – if the object is in poor condition, and if the value of the research being undertaken outweighs the level of damage which will be incurred as a result.

Non-invasive techniques enable specimens to be studied without the need to remove samples and so are completely non–destructive. These techniques remain the easiest to justify ethically and are generally favoured by museum boards for this very reason. Non-invasive studies have only become a reality with the advent of scientific knowledge which has allowed for the necessary advancements in technology to be made.

Invasive techniques were most certainly the norm in the early years of the study of Egyptology as the scientific knowledge required to underpin non-invasive methods simply did not exist. Scholars were forced to rely on autopsies of mummified humans and animals in order to answer their research questions, whilst at the same time causing the destruction of valuable archaeological material. An autopsy can only be carried out once on a specimen and it generally renders the object unfit for future display and study.

The choice of which scientific technique to use in the study of mummified remains depends on what research question is being posed. Some research questions facilitate the necessity for samples to be taken, whereas others can be answered without the need to damage the objects themselves. A thorough scientific analysis of a mummy in its true sense may use both non-invasive and invasive techniques to compile a complete picture of how the person or animal lived, died and was preserved. Such a study would utilise non-invasive techniques in the first instance to gather as much information as possible and to provide information on the best sites from which to take samples. Invasive techniques would then be used to supplement and enhance the information already gathered. The invasive techniques used would only require the extraction of miniscule samples rather than the complete autopsy of the body as would previously have been the case.

3.1 – BIOMEDICAL ANALYSIS TECHNIQUES USED IN THE STUDY OF MUMMIFIED MATERIAL

A number of biomedical research techniques, many of which were pioneered and developed for use in medicine, have been applied to the study of mummified material. In the majority of cases, these methods have been used extensively on human mummies, but as yet only minimal work has focused on animal remains. As advancements are made in the study of medicine, new techniques and methods are continually being applied to mummified material to assess their potential. This constant flow of new ideas means that the study of Egyptology is an ever-evolving discipline which is striving to find the best possible methodology for the material available.

This research project focuses on the use of radiographic techniques and it is therefore not the intention to discuss other biomedical techniques in this study. Information relating to four radiographic techniques has been included in this chapter to present the necessary supporting evidence on which this research is based. Conventional radiography (X-ray) and computed tomography (CT) have been used extensively on human remains with great success. Magnetic resonance imaging (MRI) and ultrasound (US) appear to be less suitable methods when applied to mummified material, but they have been included here with the intention of eliminating them from the research model.

Radiography (X-ray)

The use of X-rays is one of the most important scientific techniques in the study of mummified remains. It has revolutionised the study of mummies, both human and animal, and it has helped to preserve many of the artefacts held in museums worldwide as studying them by this method no longer brings about their destruction. It is a method which is totally non-invasive, resulting in mummies being studied without removing them from their coffins or bandages, but keeping them totally intact. No other technique can boast this facility, which is why radiology has become an indispensable tool in mummy studies.

Having established that radiology is generally useful, it is necessary to outline the reasons why, and the types of information that can be gathered from radiological examination. Aufderheide (2003, 381) comprehensively lists the advantages of X-ray which can be summarised as follows:-

- the presence or absence of a body or bodies,
- whether the remains are human or animal in origin,
- the sex, stature, age and possible cause of death,
- the presence of any pathological conditions and fractures,
- the presence or absence of viscera, soft tissue, artefacts and any other inclusions within the bundle ie. organic matter.

In December 1895, a German scientist, Wilhelm Röntgen, took the stand at an assembly where he presented the results of his latest research. What Röntgen had discovered was the X-ray – a discovery that was to take the world by storm, both medically and academically, and which won the scientist the coveted Nobel Prize for Physics in 1901 (Owen 2000, 16; Aufderheide 2003, 381). The following year, both Röntgen and Dr. Thurston Holland produced radiographic images of mummified humans and animals as they represented 'safe' subjects on which to experiment with this new technique. Animal remains were often chosen for these trials because of their compact size which was easily accommodated by the primitive X-ray equipment available at the time. As techniques advanced, human mummies began to be studied using this method, producing many notable papers by eminent scholars such as Flinders Petrie (1898), Elliot Smith (1912) and Moodie (1931). These early studies proved that radiology showed enormous potential in the study of mummified remains from ancient Egypt.

Conventional radiology produces two-dimensional images of three-dimensional objects. As the X-ray beam passes through the object, it is absorbed differently by the many tissue types, producing an image on the film where the transmitted X-ray beam reacts with silver crystals. The atomic number and density of the tissue determines how many of the X-rays are absorbed. Bones appear white on radiographs because their high atomic number prevents the beam from passing through creating areas of radio-density. Structures with lower atomic numbers such as muscle, soft tissues, fat and air, will allow some of the X-rays through appearing on the images as darker radio-lucent areas (Aufderheide 2003, 379). These differences are referred to as radiographic contrast. Spatial resolution refers to the difference in size between component structures which can be identified (Adams and Alsop 2008, 28-30).

There are four variable factors involved in the process of X-ray (O'Connor and O'Connor 2005, 71) –

- the length of the exposure – seconds
- the X-ray current in the tube – mA
- the voltage – kV
- the distance of the object away from the source of the X-rays/the film (film focus distance (FFD) and film object distance (FOD))

For the purposes of this research project, there were a number of factors which affected the decisions made regarding the variables listed above. Mummified material has often been treated with substances such as resin or

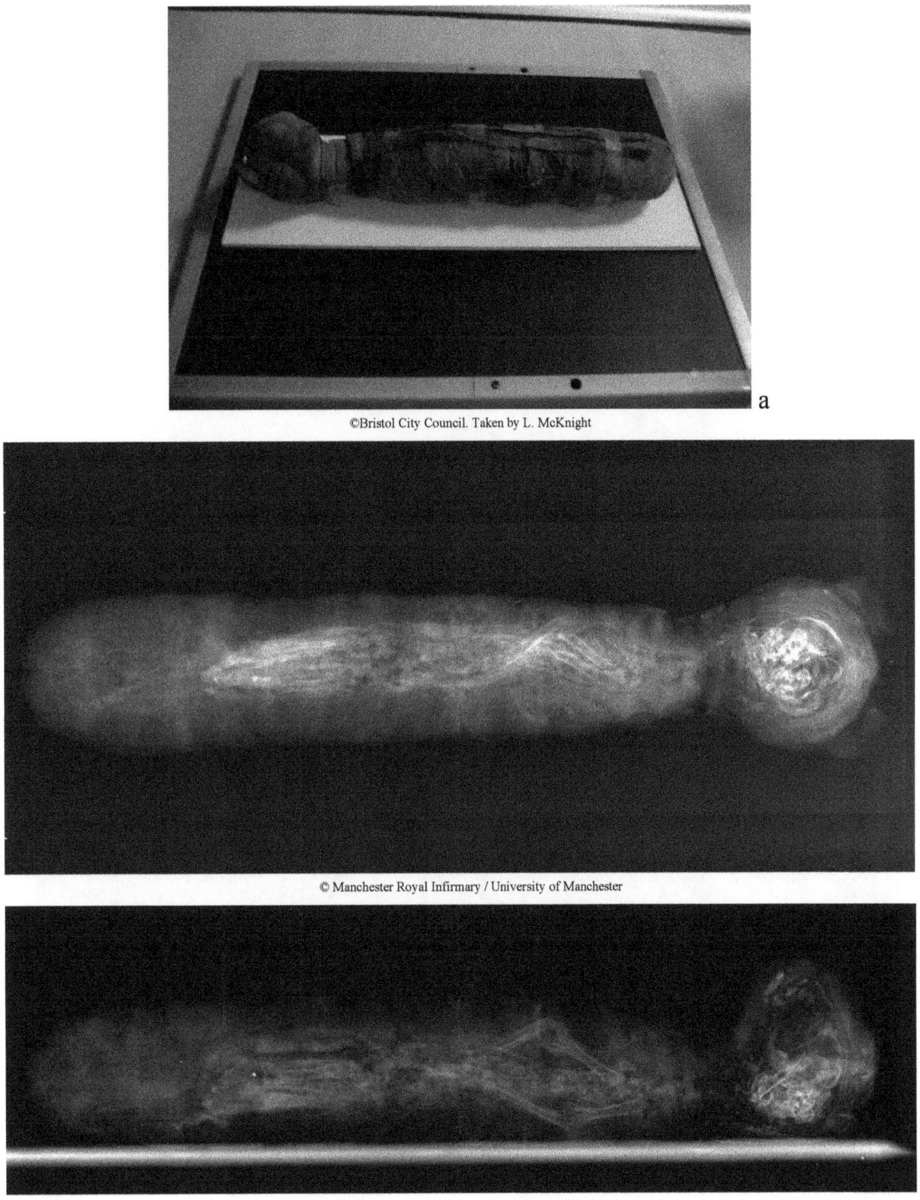

Fig. 3.1 – a) Photograph, b) AP radiograph, and c) Lat radiograph of mummy MA2.1. The tight plastic straps fixing the mummy to the display board could be responsible for causing the skeletal damage to the neck region

bitumen which can make it difficult for the X-rays to penetrate the object. Trial and error played a major role in the values chosen for the variables simply because the radiographers executing the imaging were not accustomed to dealing with desiccated ancient animal bone complicated by the presence of substances used in the mummification process. This is undoubtedly not the most scientific way of working; however it was the most successful method of achieving successful results in this case.

Radiographic images were taken in two planes – anterior posterior (AP) and lateral (Lat). AP images essentially examine the object from top to bottom, whereas Lat projections view from the side of the object (Fig. 3.1). Imaging each specimen in two planes can help to minimise the effects of magnification and superimposition by providing the researcher with two different views, therefore aiding interpretation (Fig. 3.2). In the case of specimens that were particularly 'flat' in dimension, it was not necessary to take Lat images as they would not have added any further information.

In a number of cases, it was necessary to use padding to support the specimens during radiographic study to achieve the necessary angles for imaging (Fig. 3.3). Foam blocks proved to be an effective material for this purpose. It was also necessary to radiograph some specimens in their packaging for conservation reasons (Fig. 3.4).

Medical radiography equipment, such as that used in this study, is considered to be inferior to that used in industrial applications when studying mummified material. This is because medical or veterinary radiographic films are specifically designed for imaging living patients where the radiation doses have to be kept

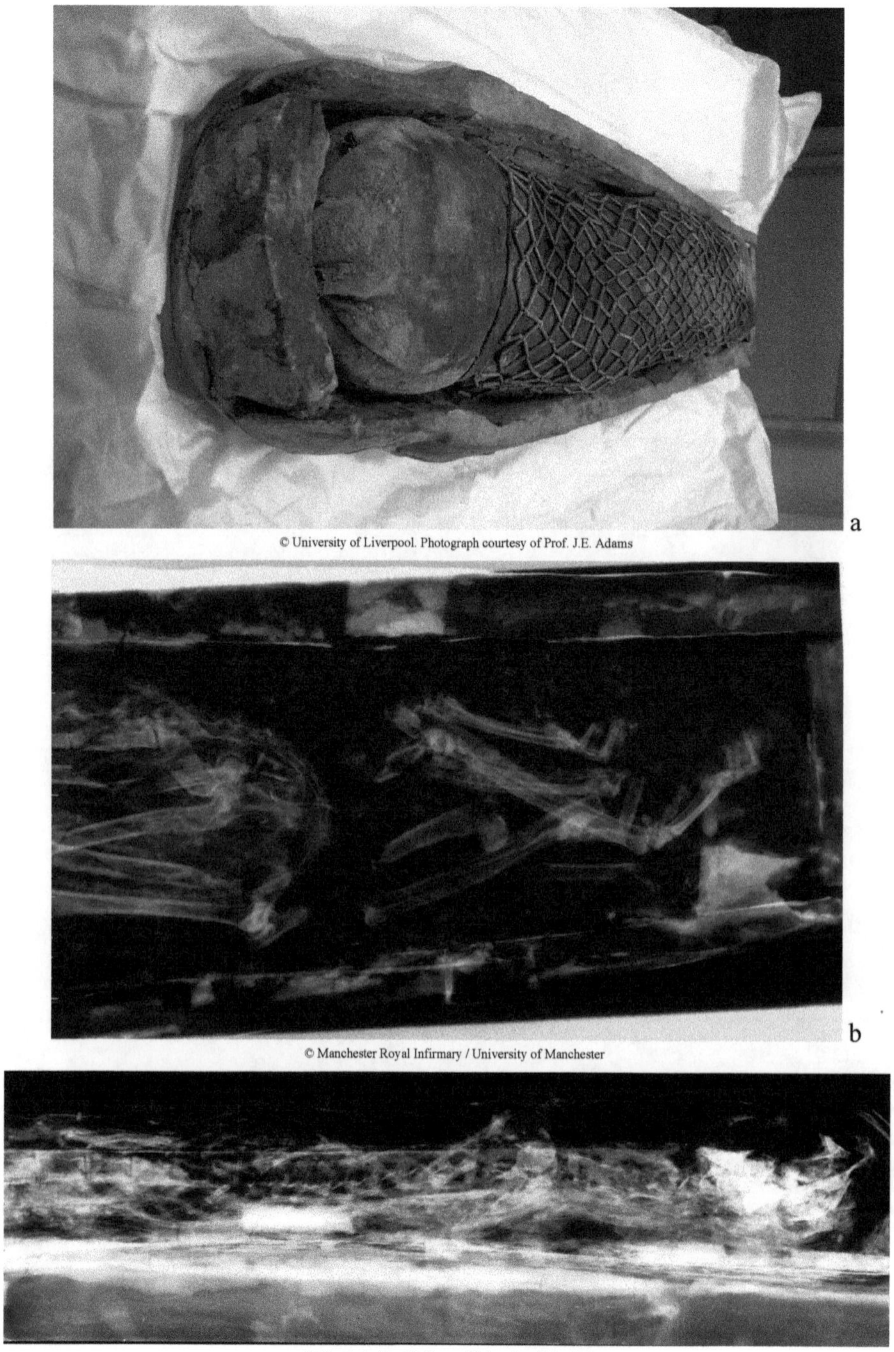

a
© University of Liverpool. Photograph courtesy of Prof. J.E. Adams

b
© Manchester Royal Infirmary / University of Manchester

c
© Manchester Royal Infirmary / University of Manchester

Fig. 3.2 – a) Photograph, b) AP radiograph, and c) Lat radiograph of mummy MA2.6. The radiographs clearly demonstrate the problems of superimposition and magnification common in conventional radiography and why it is necessary to produce images in two opposing planes to limit the effects of these phenomena. Images b) and c) are essentially of the same area of the mummy bundle, but both are needed to gain a full insight into the nature of the contents

to a minimum in order to obtain images with short exposure times. Archaeological specimens can be imaged using slower industrial films capable of withstanding greater radiation doses and longer exposure times, but which can achieve far superior image quality on ancient bone (O'Connor and O'Connor 2005, 70).

The radiographic study of the Bolton Museum Collection by the author in 1999 used industrial film (higher spatial resolution) and resulted in good quality images despite the fact that they were developed by hand in less than ideal darkroom conditions (Owen 2000). It is difficult to attempt a comparison between these images and the ones produced of the same specimens during 2005 using medical film as the method of film acquisition and development differed.

As well as cost implications, X-ray does have some disadvantages when studying mummified material which have meant that more recent technological advancements

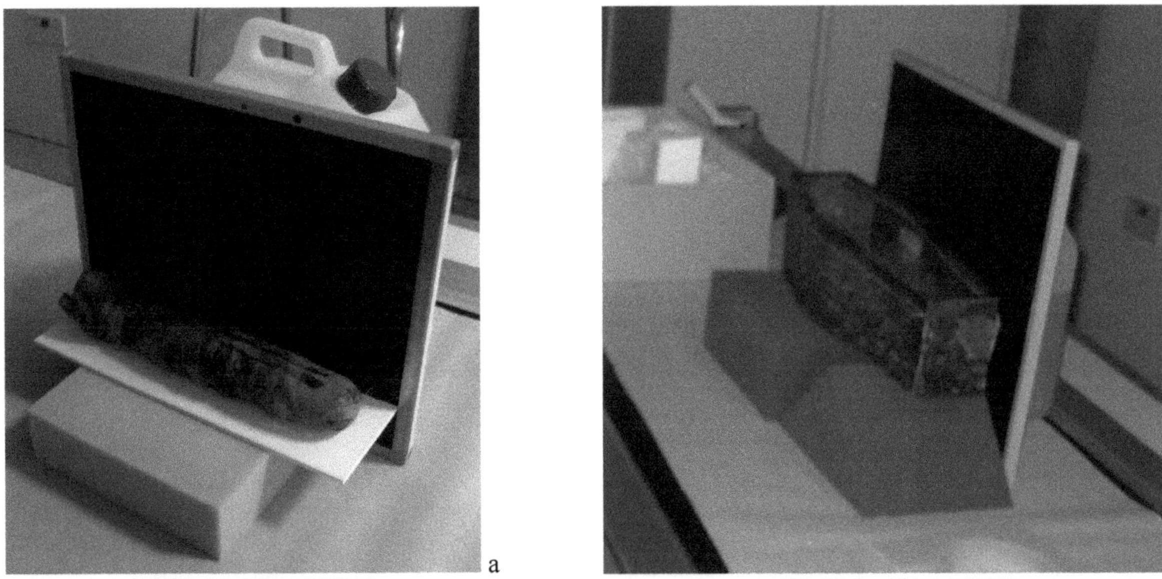

a b

Fig. 3.3 – Photographs a) and b) show how foam blocks and water-filled containers were used to support the delicate specimens during the imaging process. This was necessary as many of the specimens needed to be raised up off the bed in order to produce the best images

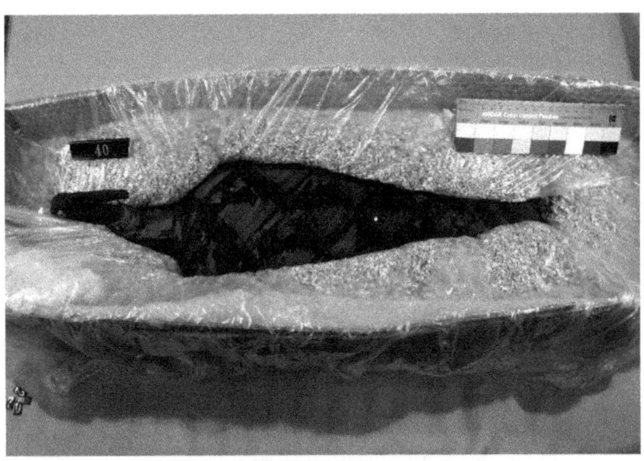

a

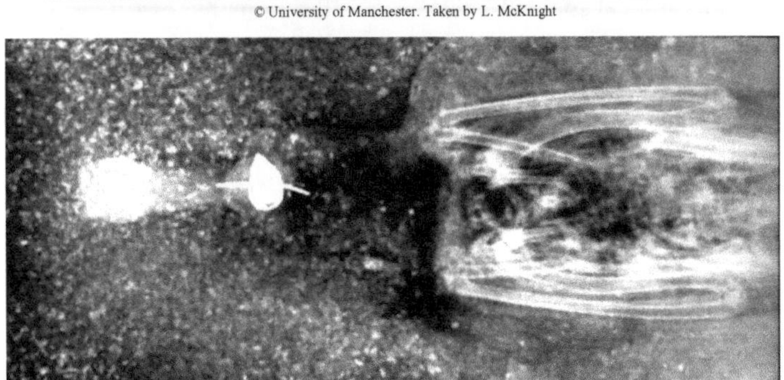

b

Fig. 3.4 – a) photograph and b) AP radiograph of mummy MA1.20 which was packed for transport and study in a box filled with vermiculite pellets. Image b) highlights the poor quality of the resulting radiographs caused by the vermiculite

have introduced more suitable methods for use on this type of material. Perhaps the major concern is that in producing a two-dimensional image, there is a large amount of superimposition caused by overlap of anatomical and other structures. This is compounded by the fact that the position of mummies cannot be changed to minimise such overlap as can be achieved with a living patient. Also the materials used in the mummification process and the desiccation of the tissues reduces radiographic contrast differences which can provide misleading results and which often facilitates further radiological investigation.

Structures which lie farther away from the X-ray film are subjected to magnification and can therefore appear larger than they actually are. O'Connor and O'Connor describe the process in the following way (2005, 71):-

> 'The X-ray beam forms a cone diverging from the point of production. The further away the film is from the source, the wider the beam and therefore the less intense the beam. This reduction in intensity follows the inverse square law, so doubling the distance reduces the intensity of the beam to a quarter and quadruples the length of the exposure. On the other hand, the closer a three-dimensional object is to the X-ray source, and the further the beam is from the perpendicular when it reaches the film, the greater will be the distortion of the two-dimensional object.'

This can be a problem in the radiography of mummified material as it becomes impossible to gain accurate measurements of the structures visible on the images. In an area of study where it is not possible to unwrap the mummy bundles and obtain direct measurements, Egyptologists have to acknowledge the issues surrounding magnification and move beyond them. The degree of magnification experienced in X-ray will be discussed in Chapter Ten. Contact radiography (where the specimen is laid directly on the film) and using a long Focus Film Distance (FFD) (100 cm or more) reduces magnification to a minimum. If this is not possible then methods are available to correct for the degree of magnification (using a metal measuring object included in the radiograph at an appropriate plane of the object).

Blurring and a general lack of sharpness on radiographic images occurs due to the scattering of radiation through the objects or substances which lie between the source and the film, and is also dependent on focal spot (FS) size (lower with fine FS of 0.6 mm). These objects could be anything from the air surrounding the object to the structures which make up the object itself. Radiographic equipment is inherently flawed as it causes a degree of blurring of structures placed within it. O'Connor and O'Connor (2005, 71) comment that:-

> 'by maximising the X-ray source-to-object distance and minimising the object-to-film distance, this source of image degradation can be largely eliminated.'

With all forms of radiological study, it is possible to create a digital record of the objects studied which can be used as a resource in the future. This is an important consideration and a major advantage in the use of this technique. Modern radiology utilises complex computer technology to digitally format the information which can then be stored, duplicated, printed and distributed where necessary. This means that the delicate objects do not need to be transported as frequently and that time spent carrying out research is kept to a minimum. It is essential that digitisation is carried out using high-quality techniques which if performed successfully can enable manipulation of the image and subsequently aid in their investigation (O'Connor and O'Connor 2005, 75-78). Aufderheide (2003, 379) summarises the importance of X-ray succinctly by saying:-

> 'plain films commonly generate considerable basic information and can be prepared for only a little more than the cost of the film'.

Computed Tomography (CT)

CT essentially became a clinical possibility in the 1970s with the advent of computer technology sophisticated enough to allow the manipulation of radiological images and the processing of large amounts of data. The initial concept was developed by J.H. Raden in 1917, but technology at this time did not permit the idea to be adopted by the scientific community and it was not until 1957 that the medical applications of the method were recorded by A.M. Cormack in South Africa (Kalender 2000, 13-14). Cormack never actually used CT; it was a scientist working for the British company EMI Ltd, G.N. Hounsfield, who was the first person to put the theory of CT into practice in 1972 (Hounsfield 1973; Ambrose and Hounsfield 1973; Hounsfield 1976). The first full body CT scan of a patient was conducted at the Atkinson Morley Hospital, London, in 1972, and by 1979 the technique had become so universally renowned, that both Cormack and Hounsfield were awarded the Nobel Prize for Medicine.

CT takes X-ray projections from multiple positions around the object and produces multiple cross-sections by detecting transmitted X-ray photons by crystal scintillation (Notman 1986). Kalender (2000, 20) describes the concept by saying that:-

> 'the principle of CT consists of measuring the spatial development of a physical quantity to be examined from different directions and to compute superimposition free images from these data'.

A full body CT scan can produce many hundreds of separate sectional images which can be used to produce a three dimensional reconstruction of the object, a method which has had important implications for disciplines such as facial reconstruction, in which it is no longer necessary to have physical access to the skull of the subject. CT has many positive attributes for the study of ancient mummified material, both human and animal. Like conventional radiography it is completely non-destructive enabling whole mummies to be studied without the need to destroy the resource.

One of the most recent displays of the breathtaking contributions of CT technology to Egyptology is the exhibition at the British Museum, London, which has literally produced a 'Virtual Mummy'. The exhibition features a complete provenanced human mummy within a sealed decorative cartonnage which has been CT scanned

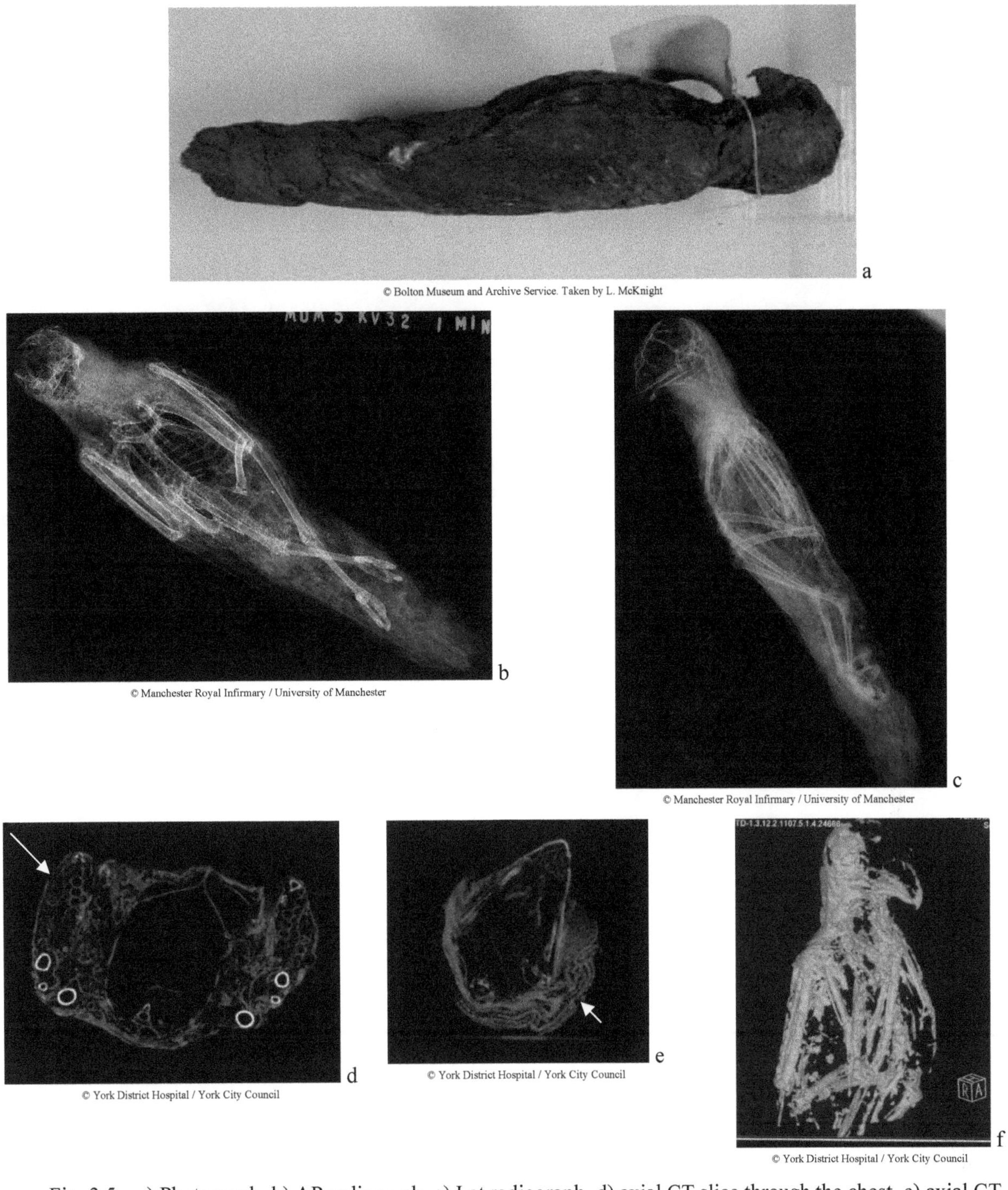

Fig. 3.5 – a) Photograph, b) AP radiograph, c) Lat radiograph, d) axial CT slice through the chest, e) axial CT slice through the skull, and f) 3D reconstruction of mummy MA1.27. Image d) clearly shows the wing bones in section and the spinal column, with the rows of circular anomalies highlighted being the wing feathers. Image e) shows the exposed beak with layers of bandages visible to one side of the skull. Image f) highlights the problems associated with 3D reformatting as the mummified soft tissue is difficult to distinguish from the skeletal remains

and the images used to create a three-dimensional replica of the person inside. Using this reconstruction, scientists and Egyptologists have been able to identify pathological conditions evident in the mummified remains and also valuable information on the way in which the body was treated after death. This is perhaps the best known example to date which has enabled such extensive research into a mummy, whilst the physical body remains sealed within its original cartonnage.

CT was the first non-invasive imaging method which enabled researchers to visualize soft tissue rather than simply skeletal material (Hoffman, Torres *et al.* 2002, 378). In mummified remains, this is extremely important as soft tissue has often been well-preserved due to the desiccation process, and it can reveal important information on pathological conditions and the general health of the individual. Pathologically, the nature of soft tissue is known to yield more physical evidence of disease than bone, so mummies present an exciting medium for this area of study. CT also offers the opportunity to visualize any cavities within the body, and to differentiate between contents which they may hold, for example, linen bundles, internal organs or mummification materials.

Until recently CT had another valuable advantage over X-ray in that the images were available digitally and could be directly processed using computer algorithms which could be reproduced to form both two- and three-dimensional images (Kalender 2000, 150). Archiving systems such as Picture Archiving and Communications Systems (PACS) and Computed Radiography (CR) allow radiographs to be viewed with high contrast resolution making the resulting quality of the images much improved (Adams and Alsop 2008, 24-25).

The sectional nature of CT lessens the effects of superimposition which is a major problem in X-ray. Manipulating the images to create reconstructions enables the researcher to view the object in three-dimensions and without overlap of structures (Fig. 3.5). In the case of mummies, the processes involved in mummification often leave visible evidence of how mummification was carried out, and the materials used. The higher contrast resolution of CT allows greater detail of structures which simply appear as soft tissues on radiographs to be identified.

> 'The benefits of CT include superior contrast and spatial resolution and the capacity to remove superimposed embalming materials from images of internal structures'
>
> (Hoffman, Torres *et al.* 2002, 378).

CT is still a relatively new imaging technique which is constantly being adapted and refined through technological developments which enable the equipment to be continually updated. The popularity of the method has been borne out of the success it has showed, encouraging institutions that have access to the required funding to install the relevant equipment. EMI created the first ever CT scanner in 1974 when there were a total of sixty scanners in the world. By the end of the 1970s, eighteen companies were marketing their own brands of CT imaging equipment, and by the year 2000, there were in excess of 30,000 scanners in the world (Kalender 2000, 15). CT equipment can cost between £ 0.5 and £ 1.5 million depending on the sophistication of the unit, but once this has been purchased, each scan costs relatively little to produce. The only problem with the rate of progress in this field is that machines become outdated and defunct quickly as they are superseded by newer technology.

Developments in CT technology include the ability to image thinner sections (as small as 0.6 mm) and the introduction of spiral multi-detector (MDCT) scanners which are able to interpolate the spaces between the sections which enables 3D volume imaging. The scanning process is becoming increasingly faster with a human torso now taking around twenty seconds to scan completely (Adams and Alsop 2008, 26). The original CT scanners acquired 2D images of 5-10 mm in thickness and took about twenty seconds each to acquire.

The radiation doses needed to perform both X-ray and CT vary depending on the quality of image required and the material which is being imaged. The radiological investigations of the animal mummies carried out at the Manchester Royal Infirmary in January 2005 incorporated radiographic images taken at 100 kV and 10 mA largely because they are so small and compact, therefore not requiring large radiation doses. This dose would be acceptable on a living subject, but often it is necessary to use higher doses to achieve adequate images of mummified material. Mummified tissue has the advantage over the living patient as the dosage is largely irrelevant although it may affect the quality of tissue samples should DNA analysis be required in the future. It is therefore advisable to take any tissue samples from the mummies before radiological examinations are carried out. This is not always easy as radiology may be necessary to assess the presence and position of structures from which the samples will optimally be extracted.

Magnetic Resonance Imaging and Ultrasound (MRI) and Ultrasound (US)

Many scientists have witnessed the amazing impact that CT has had on medicine and Egyptology and have been inspired to test the more recent radiological techniques to have been developed: MRI and US. These techniques rely on the detection of water (US) or hydrogen protons (MRI) which are found in liquids and some fatty materials. Mummified material is by its very nature dehydrated and so not the perfect candidate for scanning using these methods (Adams and Alsop 2008, 26-28). However, they may still have useful applications in this field.

Notman attempted MRI on mummified material in 1995 and found that even though it held no direct use in its study, it could be an aid in experimental studies to chart the extent of the dehydration throughout the process. Another possible area in which these methods could be useful is on rehydrated mummified material as they have proven to be effective on preserved bog bodies such as Lindow Man (Aufderheide 2003, 380).

Chapter Four
THE COLLECTIONS

4.0 – THE MATERIAL

The mummified specimens studied during the course of this research project were made available for radiographic analysis by their holding institutions following a direct mailing which hoped to establish contact and assess viability for further study. The response to this mailing surpassed expectations and revealed a large number of animal mummies around the United Kingdom, most of which had never been studied in any detail. The resulting study group contained 127 mummies in total. This was felt to be a significant group of sufficient size to enable the radiographic techniques selected to be tested adequately. The majority of the mummies made available for study came from four museums:-

- Bolton Museum and Art Gallery
- The Department of Archaeology, University of Liverpool
- Bristol Museum and Art Gallery
- The Manchester Museum

A small number of mummies were donated for study by:-

- Stockport Metropolitan Borough Council
- Private collectors

4.1 – THE CATALOGUING SYSTEM

The disparate nature of the documentary records held by the museums concerned and the haphazard numbering systems which have evolved through centuries of manual records and individual acquisitions prompted the formulation of a cataloguing system specifically designed for this project. This system was created to aid the interpretation of the reader by separating the collection into species groups which it was hoped would lead to a more comprehensive study of the practice of animal mummification as a whole.

The mummies were divided into the following groups:-

1. Birds
2. Cats
3. Dogs
4. Fish
5. Crocodiles
6. Other Animals
7. Pseudo-Mummies
8. Other Varieties

Specimens belonging to each of the above groups were then allocated an individual number. For example, the bird mummies within the study group were allocated numbers between 1.1 and 1.35. All of the new numbers shared the prefix MA to identify them as Manchester Animals (animal mummies studied in the Manchester project). This system removed the confusion and ambiguity surrounding the original museum records and created a coherent system for the project. It was hoped that this system, if showed to be successful, could be used to catalogue human mummies in further studies using the prefix 'MH'.

A comprehensive key to be used in conjunction with this numbering system has been included in the Appendix. The purpose of the key is to collate the various acquisition and museum numbers which have been associated with the specimens through time to enable the reader and future researchers to be able to easily locate the mummies concerned at their holding institution should the need arise.

The Appendix also contains a written catalogue containing all the known information relating to each individual specimen. This includes details of the museum where the mummy is located, museum documentation, external observations, and the radiographic findings. The sheer volume of information relating to the mummies prompted the decision to include a full catalogue in the Appendix with relevant material being selected to demonstrate key points within the main body of the text.

It should be noted that this numbering system was devised following the completion of the radiographic analysis and the classifications are therefore based solely on the results of this. For the purposes of this numbering system, the specimens classified as pseudo-mummies are those which were found to contain no skeletal material identifiable as being of animal origin. I would usually prefer to define pseudo-mummies as being any specimen which is found to be something other than what it was made to resemble and this can often means that they do contain some animal material. For the purposes of this study and to enable a more coherent discussion, the decision was taken to use this alternative classification for ease of interpretation.

Chapter Five
ANIMAL TYPES

5.0 – THE RESULTS EXPLAINED

For the purposes of this research project, I have decided to display the complete results of the radiographic analysis in the form of a catalogue which is included in the Appendix. Selected photographs, radiographs and CT images have been included on the accompanying CD. This decision was made due to the large volume of information gathered during the course of the project which made it impractical to include it in the body of the text.

A complete results table appears at the start of the catalogue although this is intended only as a summary and readers should consult the Appendix for the extended results. Specific specimens will be described in more detail in later chapters.

Due to the volume of information gathered from the study collection, there are five separate chapters in this publication dedicated to the results of the radiographic analysis. The content of the chapters has been categorised in the hope that the reader can gain a full understanding of the material studied. The results chapters are as follows:-

- Chapter Five – Animal Types – species identification, age, sex, body cavity contents, pathological conditions, cause of death.
- Chapter Six – The Mummification Process – external appearance, body shape, evisceration, position of the body, packing of the body cavities, wrapping, inclusions within the mummy bundle.
- Chapter Seven – Pseudo-mummies – mummies containing part of an animal skeleton or no animal material at all.
- Chapter Eight – Other Varieties – mummies which produced interesting and surprising radiographic images.
- Chapter Nine – Case Studies

This chapter is intended to present information gathered from the radiographic analysis which relates to the animals themselves. The focus of the discussion will relate to votive mummies as it is believed that the majority of the specimens studied during this research project belong to this category. They were living creatures who were intentionally bred in captivity, raised and then slaughtered to provide votive objects purchased by temple pilgrims as a means of communicating with the gods. With the advent of modern technological accomplishments, these animals are able to convey secrets about their lives and deaths to Egyptologists and scientists.

A small number of more elaborately mummified and decorated animals may represent pet or cult animals, and as such were treated differently by the ancient Egyptians. It is not possible to state with certainty whether this theory is correct as the specimens in question have little in the way of provenance or dating information, although they undoubtedly stand out from the others in terms of external appearance alone.

The Research Questions

The types of animals identified in this project will be discussed individually with regards to a number of research questions which it was hoped would be addressed through the radiographic analysis. These research questions are introduced to the reader in the following paragraphs to outline the potential issues which can arise during research into this subject.

Identification of Species

The identification of the species can be attempted through the visual inspection of the external features of the mummy bundle, but radiography provides an insight into the interior in a non-destructive manner. X-ray is the preliminary method of scientific analysis that should be applied to any mummified material, whether it is human or animal. X-ray is the cheapest, fastest and completely non-destructive method of seeing the contents of the mummy bundle; however it can pose some issues which can prove to be problematic for species identification.

The major concerns surrounding the use of radiography are that the method causes features within the mummy bundle to appear superimposed and distorted on the resulting image. This happens because X-ray is capturing a two-dimensional image of a three-dimensional object. As many of the techniques used to determine species rely on the identification of diagnostic skeletal features, any superimposition and distortion displayed on the radiograph can make the process difficult.

During the course of this research project it was necessary to seek expert advice regarding the identification of the animals in the study group. This is a specialised area and one which undoubtedly lies beyond my expertise. The individual specialists or institutions who offered their support and advice in this area are discussed below in the species sub-sections.

When studying ancient animals, an added difficulty surrounds comparing modern species with their ancient counterparts. Over the centuries since the time of the ancient Egyptian civilisation, the fauna of the country has changed dramatically as species have migrated away from the Nile valley in search of new habitats. Combined with this is the fact that many species which are still present in modern-day Egypt have evolved to become domesticated and have consequently altered skeletally. Usually the scale of this evolution is minimal although it is important to acknowledge the possibility of it occurring in case it has a bearing on the results of the analysis.

It should also be remembered that the comparative collections available in institutions around the United Kingdom generally contain the remains of English and, in some cases, European animals. Although there are a number of problems concerning the use of this type of comparative skeletal collection, they are often the only means of aiding identification that researchers have available to them when working with mummified animal material.

When attempting to identify the other species within the study group it was not possible to use comparative skeletal material because of the unavailability of such collections. As an alternative method of identification, copies of the radiographic images were studied in depth by veterinary specialists whose expertise enabled them to identify diagnostic skeletal features. These specialists include Henry McGhie (Manchester Museum), Wim Van Neer (Museum of Central Africa, Belgium), Mark Beech (Department of Archaeology, University of York), and staff at the Environmental Archaeology Unit (University of York).

Ageing and Sexing of the Animal

Initial estimations of the age of an animal can be made through visual observations of the size of the skeleton itself; however, this is not without its problems. There can be a wide margin of size difference between animals of the same species and there are also concerns regarding sexual and regional dimorphism.

The major method of establishing the age of an animal through the analysis of radiographic images concerns the study of the extent of epiphyseal fusion. The epiphyses are the plates at the end of the major long bones of the skeleton which are separated during youth by layers of cartilage. As enchondral ossification occurs, the bones increase in size and shape, leading to the growth and development of the skeleton (Adams and Alsop 2008, 35). The fusion of these plates with the long bones takes place at certain points during life which have been identified through veterinary and medical research. Radiographs provide a 'snap-shot' of a creature's skeleton at the age of death and the identification of any visible unfused epiphyses can suggest an 'age window' during which the animal died.

In principal, the ageing of an animal through the study of epiphyseal fusion is impressive; however this too is not without problems. The long-standing issues of distortion and superimposition of features on radiographs can make the identification of the epiphyses very difficult and in some cases impossible. In this research project the animals have simply been classified as juvenile or adult as it was felt that any further age distinction would have been speculative.

A secondary method used to help determine age at death is the degree of dental eruption. This method, used in the study of both human and animal remains, relies on known patterns of tooth development. In the study of archaeological remains, it can be difficult to assess dental eruption in specimens where extensive damage has occurred post-mortem.

Establishing the sex of a mummified animal can be extremely difficult. Essentially the aim is to determine the sex using skeletal indicators and measurements although these can be hard to distinguish on radiographs. For these reasons sexing of the animal within this study was not attempted.

Body Cavity Contents

Radiography can provide evidence of contents within the stomach or cranial cavity. This could be either a naturally occurring substance such as dehydrated brain or orbital matter or evidence of the last meal consumed prior to death. Evidence of intentional packing by the embalmers will be discussed in section 6.0 as this relates more to the treatment of the body after death rather than the remains themselves. The presence of dehydrated matter suggests that the body was mummified before the natural processes of decay began to act upon them. The brain is usually the first organ to be affected by autolysis after death, so the presence of dehydrated brain matter in the cranial cavity suggests that mummification took place immediately following the point of death.

Evidence of the remains of the internal organs in the form of desiccated tissue indicates that it was not common practice for these to be removed as was generally the case in human mummies and high status animal burials. The practice of evisceration will be discussed in section 6.0.

Evidence of Pathological Conditions

The subject of pathological conditions suffered by animals that were mummified is something of a double-edged sword, with a variety of issues affecting their survival and interpretation.

1. Skeletal pathology is difficult to find simply because the majority of conditions do not leave lasting evidence in the skeleton (Owen 2001, 98). Soft-tissue is generally the target for disease, but as the mummification process affects the biological constitution of the tissue, it can be almost impossible to find evidence of pathological conditions. Histological testing is able to use tissue samples to test chemically and histologically for the presence of disease. These methods are fraught with problems as they rely on the particular sample being tested containing persisting evidence of the condition.

2. The scientific methods chosen to study the mummified tissue are not always compatible with the hopes of finding evidence of pathological conditions. As mentioned above, histology and related techniques used in modern medicine to confirm the diagnosis of disease require invasive methods to remove samples which, for ethical reasons, is not always deemed possible. Radiography, although totally non-invasive, is not necessarily the most effective technique for identifying evidence of disease. Skeletal features are well demonstrated on radiographic films; however, they are not indicators of disease but just demonstrate trauma (which may be a by-product of disease if it occurs as a result of weakness). Anomalies affecting the bones tend to be fractures or lines of arrested development (Harris Lines).

3. It is likely that the animals specially bred and raised in captivity by the ancient Egyptians for use in the votive mummy industry were affected by problems associated with interbreeding and genetic abnormalities. Little is known about the breeding programmes in place at the temple complexes. It seems reasonable to suggest that interbreeding did occur on some level within the animal population. Skeletal deformities indicative of the overcrowded conditions in which the animals were housed are also to be expected (Owen 2001, 98). Evidence of the diet and state of nutrition of the animals can be assessed through analysis of the physical remains, as signs of malnutrition (either through poor care or as a result of disease) can manifest themselves in the skeleton. Animals kept indoors are also likely to have suffered from rickets caused by lack of exposure to natural light.

4. The animals bred to supply the votive industry appear to have been allowed to live short lives. This is logical from a 'business' point of view as it would not be economically viable to continue caring for an animal which would not ultimately translate into a greater profit. The trend appears to be that the animals were allowed to live until the point at which they bore resemblance to a fully-grown animal. This limits the occurrence of age-related conditions (Owen 2001, 98).

5. When researching the evidence for pathology in a collection of material, it is important to remember that the archaeological material available is only a small proportion of the whole. It is important to consider whether the collection available for study is indicative of the whole, and on this point it is possible to argue both cases. Ultimately, the mummified animal material that is available for research is minimal. Millions of animals were mummified in ancient Egypt and Egyptologists are limited in the main to those specimens which have become part of museum collections around the world and the material remaining in situ at the animal necropolises such as Saqqara and Tuna el Gebel. Some of these specimens are either not available for study or cannot be transported, further reducing the number of viable mummies. On the other hand, this material is all that is available so it is necessary to make comments and provide information based on this collection. The extent to which this mirrors the animal mummy population in general is unknown.

There is little documentary and physical evidence available from ancient Egypt to suggest how veterinary medicine was practiced. The Kahun Veterinary Papyrus is the only surviving document relating to this and it was produced over one thousand years prior to the start of animal mummification for votive purposes (Quirke and Collier 2004; Quirke 2007). It is therefore necessary to make some assumptions regarding the care that the animals would have received. It seems reasonable to assume that any visible impediment suffered by an animal would have caused it to be sacrificed sooner to prevent unnecessary suffering. It is unlikely that great care would have been taken over administering treatment as the animals were raised specifically for the purposes of slaughter.

Evidence of pathological conditions in mummified animals is relatively rare. This does not necessarily mean that the animals were free from disease. Egyptologists and scientists rely heavily on how the physical archaeological remains can provide information on the lives of the animals concerned through the application of biomedical techniques. It is important to be mindful of the fact that just because pathological evidence is difficult to identify does not mean that it did not exist (Siegel 1976, 355). A variety of pathological conditions would undoubtedly have been suffered by ancient Egyptian animals, yet their physical remains are unable to convey the full extent of these. It is only possible for

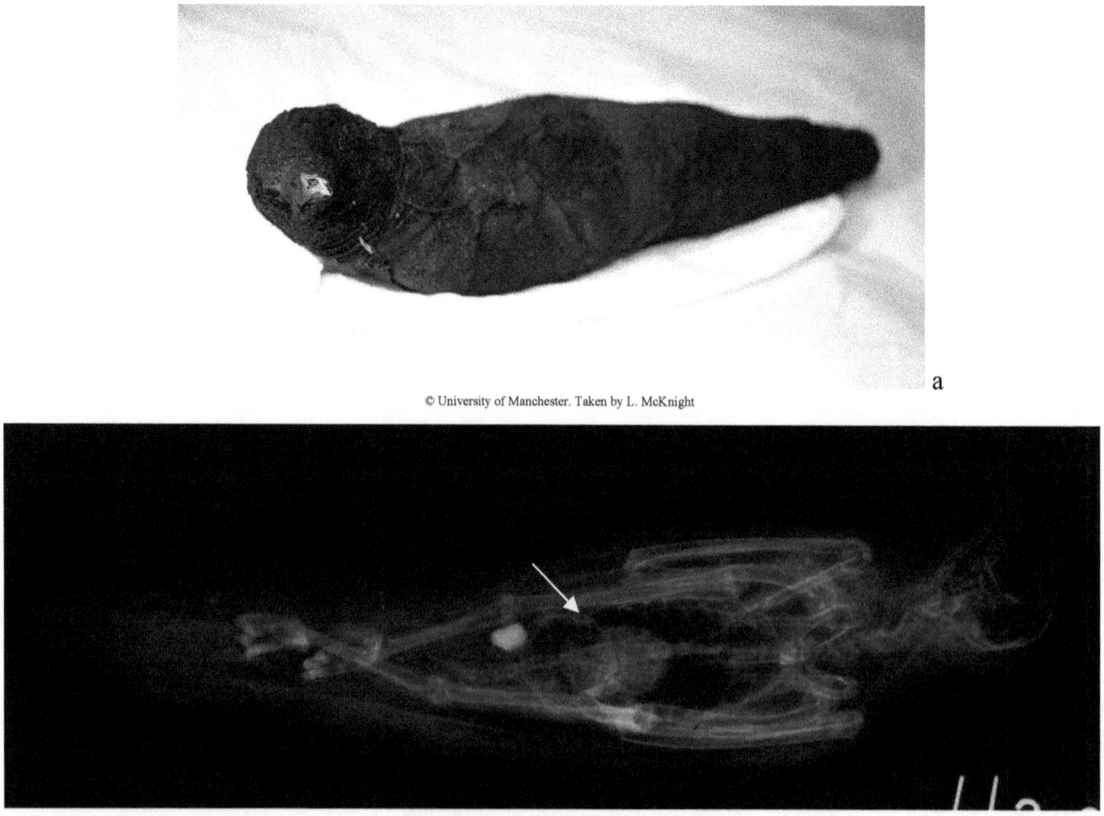

a

© University of Manchester. Taken by L. McKnight

b

© Manchester Royal Infirmary / University of Manchester

Fig. 5.1 – a) Photograph and b) AP radiograph of mummy MA1.15 showing a mass in the abdominal cavity

Egyptologists to say that little evidence was found, not that it does not exist.

Evidence of Cause of Death

Establishing the cause of death of the animals is extremely difficult and the majority of comments made are speculative. Radiography is able to demonstrate physical abnormalities of the skeleton which can be used to infer cause of death. The problems of superimposition and overlapping can make cause of death hypotheses misleading by creating a distorted image of the skeleton. Ancient material is examined through modern eyes and attempts are made to suggest the rationale behind human behaviour. As such, speculations in this field often relate to how modern-day people would approach the situation as it was seen by the ancient Egyptians.

The range of animals mummified by the ancient Egyptians was immense and each particular species would require a different method of slaughter. This depended on the species itself and the resources available. The way in which the ancient Egyptians perceived a species is likely to have had a bearing on the method used.

The results of the radiographic analysis are given below. Separate sub-chapters have been assigned to each of the different species identified to enable the reader to select relevant information more easily and to provide a more comprehensive review of the radiographic findings.

5.1 – BIRDS

Species Identification

The majority of the birds were identified to species by Dr. Joanne Cooper at the Natural History Museum's Bird Group, Tring, Hertfordshire. Comparative skeletal collections were used to achieve what is believed to be the most comprehensive set of results that could have been hoped for. Although this is not a fail-safe method when comparing bones to those on radiographs, it is one of the easiest methods to employ on mummified birds as it is relatively easy to place the bone on top of the image and make comparisons. In some cases it was possible to achieve specific identifications whereas others were simply classified as cf. falcon or cf. hawk due to the enormous similarities between sub-species.

Ageing and Sexing

None of the specimens studied displayed evidence to suggest that the birds were juvenile at the time of death. There were no cases where incomplete epiphyseal fusion could be identified which would have been one of the primary skeletal indicators to suggest that they were juvenile. All of the specimens which were found to contain a complete skeleton have been recorded as mature in the light of these findings.

It was not possible to attempt to assess the sex of any of the mummified birds studied.

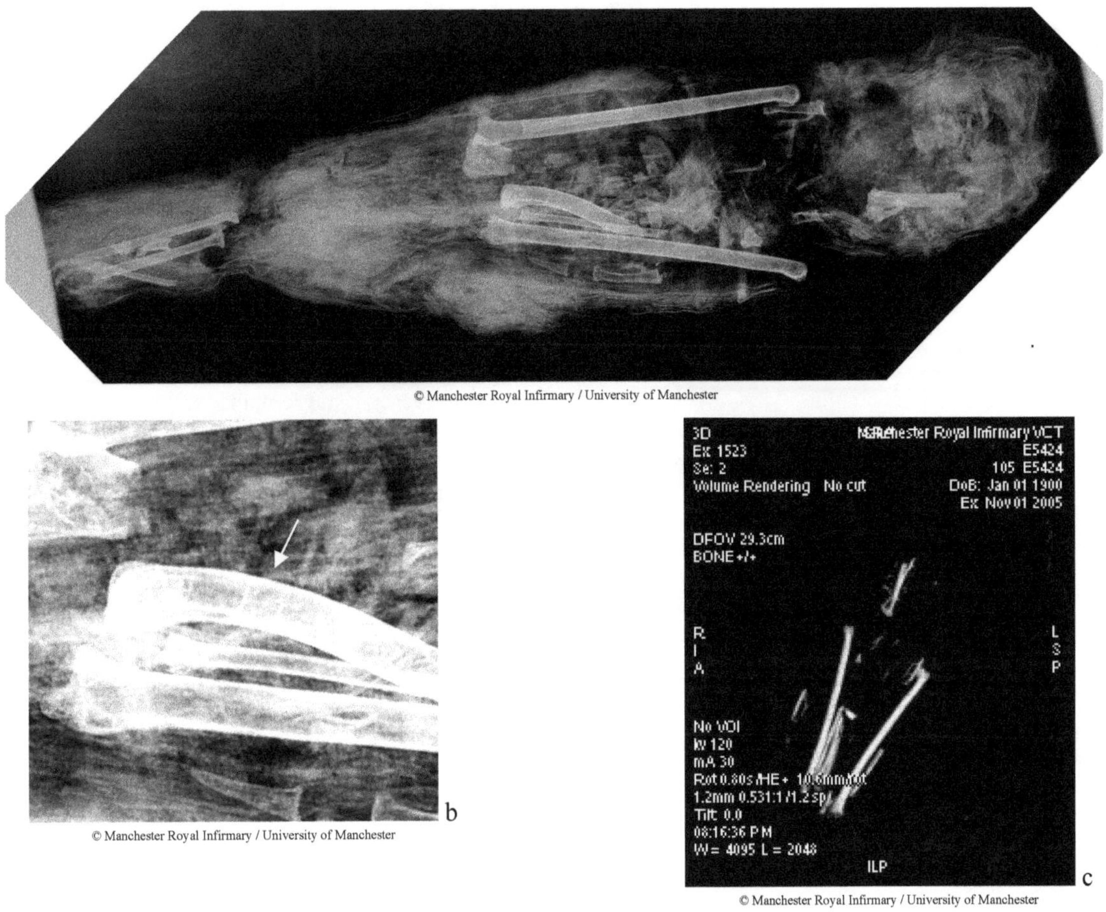

Fig. 5.2 – a) AP radiograph, b) region of interest and c) volume rendering image of the skeletal remains within mummy MA1.6. The mummy bundle contains the fragments of a juvenile ibis skeleton along with part of a falcon both of which are disarticulated and jumbled. Image b) highlights an area on the humerus of the ibis which shows the presence of Harris lines indicating that the bird suffered from a period of ill-health or malnutrition which temporarily halted the development of the bone

Body Cavity Contents

Contents are visible within the body cavities of a large proportion of the bird mummies indicating that the internal organs were left in situ during the mummification process (Fig. 5.1). These contents vary between desiccated brain matter, thoracic and stomach contents and dehydrated orbital matter.

Evidence of Pathological Conditions

Interestingly, the bird mummies revealed the most evidence for pathological conditions of any of the species studied. There was evidence of lines of attested growth in a number of specimens showing that the animals suffered from periods of ill health or malnutrition during their lifetimes, but from which they recovered (Fig. 5.2). This finding may provide evidence to suggest that the birds bred in captivity and kept within the confines of the temple complexes may not have received the best possible levels of care, and possibly that they were fed on a diet which did not contain adequate nourishment. The alternative explanation could be that the birds were not able to exercise freely within their holding pens and as such failed to thrive physically. These theories are merely speculation as there is no documentary evidence to support them; however it seems reasonable to suggest that with the sheer volume of animals being raised at any one time, the welfare of the animals would inevitably have suffered to some degree.

The single outstanding specimen in the collection is that of MA1.25 which consists of two complete disarticulated ibis femora (*Threskiornis aethiopicus*) presented unwrapped in a cardboard box. One of the femora displays a well-healed fracture mid-way down the shaft which has fused at an angle of ninety degrees. It has been estimated that a fracture such as this would take up to six months to heal properly as demonstrated by the specimen. What is even more interesting is that the bird in question must have been severely impeded by the fracture and the deformity of the limb which occurred as a consequence. It would appear that the ibis received no treatment for the debilitating injury and must have struggled to lead a normal life whilst suffering a considerable degree of pain. A simple wooden splint would have immobilised the bone allowing it to heal in a more natural position, but this evidently was not applied. One must wonder why the temple priests allowed the bird to survive with such an injury if they knew that it was destined to be sacrificed eventually as a votive offering. On the other hand, if ibises were kept as cult animals at the temple complexes one would expect them to be allowed to live until their natural deaths, but this does not explain why the animal received no treatment for its injury.

The radiographs of two of the bird mummies studied (MA1.6 and MA1.14) show evidence of Harris Lines in the wing bones and the humerus suggesting that both animals suffered periods of ill-health or malnutrition from which they recovered (Fig. 5.2). Harris Lines occur when the formation of healthy bone is retarded temporarily leading to diagnostic bone anomalies. They are useful as they demonstrate that animals displaying these characteristic lines must have recovered from the condition and gone on to lead healthy lives.

The veterinary treatment of the animals bred and raised at the temple complexes is a topic with little or no archaeological support other than the evidence from the animals themselves. The only surviving document referring to such treatment is the Kahun Veterinary Papyrus which dates from the Late Middle Kingdom, around 1200 years before the rise in the popularity of votive animal mummification and the date from which the vast majority of the study group are assumed to belong (Quirke and Collier 2004; Quirke 2007).

We can assume perhaps that the animals would have received treatment for some injuries and conditions from the temple priests who may have had a basic understanding of diagnosis and treatment of ailments. It is unknown as to whether trained veterinarians formed part of the temple staff. The nature of the work in which they were involved suggests that the animals were seen as a precious commodity which should be protected from disease which would have spread rapidly throughout the bird colonies. Their status as a commodity probably ensured that they were not treated as sacred beings in the same way that the cult animals were. They were short-term investments which must be protected until the time came for them to be dispatched for the manufacture of animal mummies.

Evidence of Cause of Death

The mummified birds within the study collection yielded very little information to suggest a possible cause of death. None of the thirty-five specimens showed physical signs of trauma which could be conclusively attributed to peri-mortem events. A large majority of them displayed evidence of bone fractures, although it was not possible to suggest whether these occurred at the time of death or shortly afterwards. It seems reasonable to suggest that rough treatment on the behalf of the embalmers could have caused further skeletal damage to the fragile animals during their initial treatment which may not manifest itself clearly as such on the radiographic images.

Birds are incredibly fragile creatures with a delicate bone structure which could easily display evidence of trauma leading to death. The most effective method of killing a bird with the minimum of effort would be to strike it briskly against a solid object. This process would lead to evidence of bone fractures occurring as a result of the impact. Another possible method of dispatch is through the use of poisonous gas; but evidence for this from mummified remains is impossible and no archaeological evidence of this practice has been found to date. Alternatively, live birds could have been placed in airtight containers, thus causing death through lack of oxygen. Live birds could also have been held by the legs and dipped straight into hot molten resin or bitumen which would have caused instantaneous death whilst helping to preserve the remains at the same time.

5.2 – CATS

Species Identification

The twenty-one cat specimens studied in this project appear to have belonged to the two species which were common in ancient Egypt: *Felis silvestris libyca* and *Felis maniculata.* Distinguishing between the two breeds was only attempted in cases where the skeletal remains were of a complete enough nature to allow for an assessment to be made.

Ageing and Sexing

In total, ten of the twenty-one cat mummies studied have been classified as juvenile based on an assessment of skeletal size, eruption of the secondary dentition and the appearance of unfused epiphyses which demonstrate that the long bones had not fully matured at the point of death. A number of the specimens did not contain complete skeletons and as such an assessment of age was difficult. Those specimens which did not show signs of incomplete epiphyseal fusion have been described as mature for the purposes of this study.

It was not possible to ascertain the sex of any of the mummified cats for a number of reasons. Seven of the specimens consisted only of the skulls and associated vertebrae which were not deemed to possess any of the diagnostic features necessary to attempt sexing.

The radiographic study of mummy MA2.2 identified a number of diagnostic features consistent with the animal being juvenile (Fig. 5.3). A number of unfused epiphyses are visible on the limb bones and the secondary dentition is present high in the tooth sockets. Mummy MA2.4 (Fig. 5.4), MA2.9, MA2.18 and MA2.22 (Fig. 5.6) also show signs of incomplete epiphyseal fusion. Mummy MA2.3 has been recorded as being juvenile due to the tiny proportions of the skeleton. It was not possible in this case to identify the stage of dental eruption using the radiographs.

Body Cavity Contents

It was possible using radiography to identify residue in the body cavities of a number of the mummified cats

a

© Bristol City Council. Taken by L. McKnight

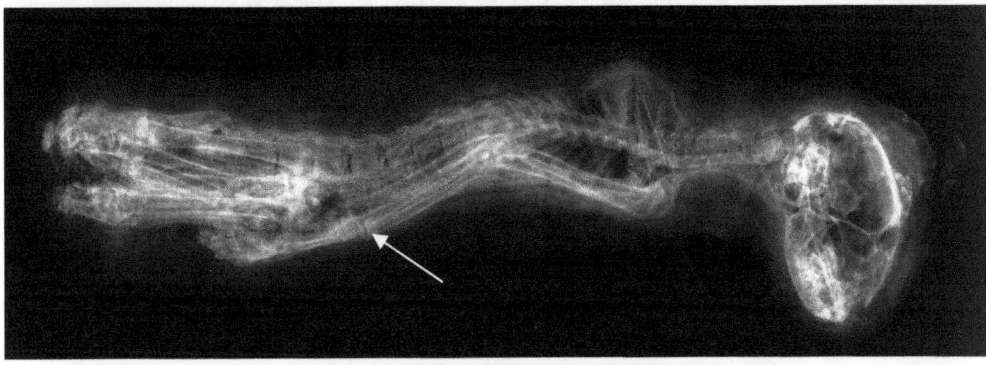

b

© Manchester Royal Infirmary / University of Manchester

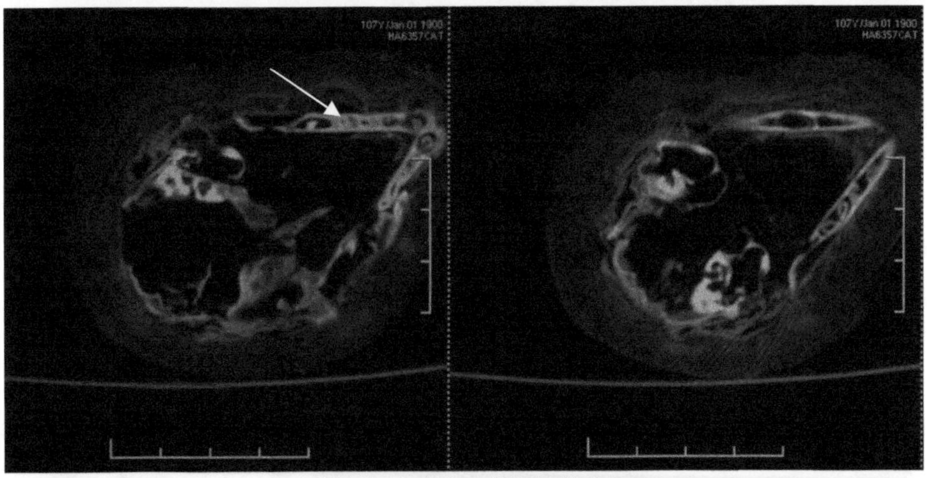

c

© Manchester Royal Infirmary / University of Manchester

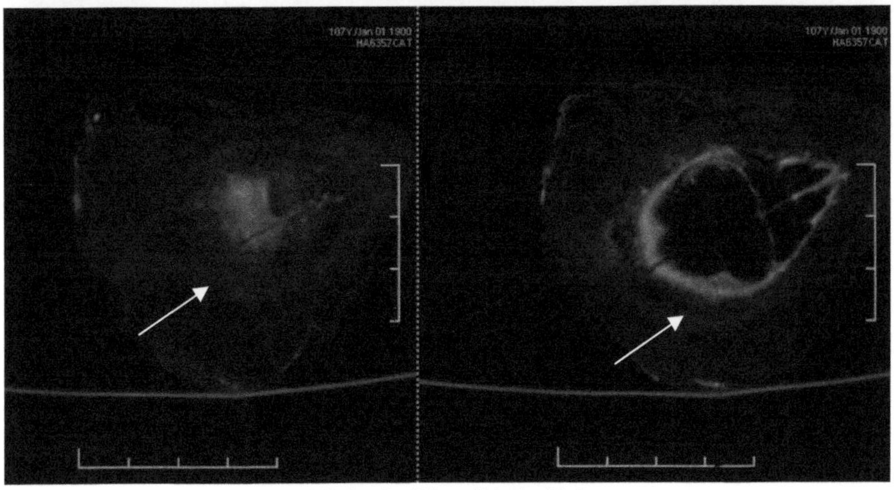

d

© Manchester Royal Infirmary / University of Manchester

Fig. 5.3 – a) Photograph, b) Lateral radiograph and c) and d) CT axial slices of mummy MA2.2. Image b) shows the presence of unfused epiphises on the limb bones indicating that the animal was a juvenile at the time of death. Image c) shows a slice taken through the skull of the cat depicting the mandible which contains unerupted teeth. Image d) shows the vault of the skull and the appearance of an unfused cranial suture, again indicating that the cat had not reached adulthood

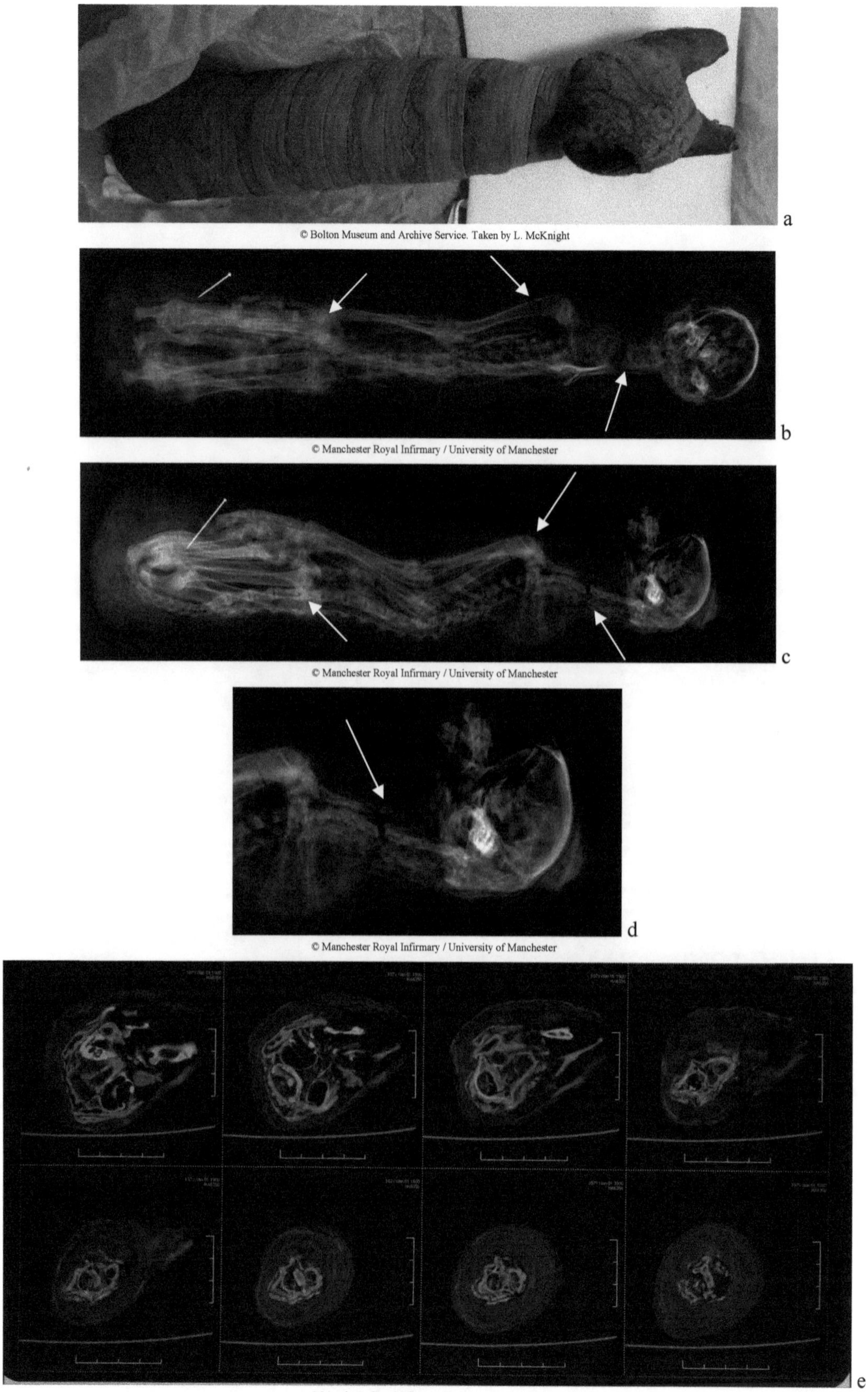

Fig. 5.4 – a) Photograph, b) AP radiograph, c) Lateral radiograph, d) region of interest and e) CT axial sections of mummy MA2.4. The markers on the radiographs show the presence of unfused epiphses on the limb bones and post-mortem fractures of the humerus and the neck. The metallic anomaly is a pin used during conservation. Image d) clearly shows the fracture to the neck; however the absence of any visible impact on the surrounding structures indicates that the fracture occurred after death, and possibly as a result of rough handling during the mummification process. Image e) shows the CT images of the base of the skull and cervical spine

studied. In the majority of cases, this appears to have been brain residue resulting from the dehydration caused by the mummification process. The isolated feline skulls in particular revealed the presence of dried brain and orbital matter possibly as the physical remains themselves are slightly larger in proportion and many are unwrapped which enabled clearer images to be gathered.

The radiographic analysis of mummy MA2.18 (Fig. 9.2.6) showed that there was brain matter present in the skull vault. The use of CT showed that the brain residue had settled towards the rear of the skull indicating the position in which the mummy bundle had been allowed to rest following mummification.

Evidence of Pathological Conditions

No evidence of pathological conditions was identified in any of the twenty-one cats studied. A number of the specimens showed evidence of bone fractures, but all were attributed to post-mortem events due to the visible bone margins which indicated that the fractures were caused after death. Many of these fractures are thought to be the result of careless handling and storage following the desiccation process when the remains would have been very fragile.

Evidence of Cause of Death

Previous research on mummified cats has suggested that they were commonly killed by having their necks broken, leading to displacement of the cervical vertebrae (Armitage and Clutton-Brock 1981; Malek 1993; Owen 2001; Ikram and Iskander 2002). This would have been an appropriate method of dispatching the animals as hastily as possible and so seems probable. Many mummified cats show signs of displaced cervical vertebrae which are indicative of a broken neck. It is difficult to determine if trauma occurred pre- or post-mortem. This method of killing may also have been employed for dogs and monkeys.

No evidence could be found to suggest that the animals studied had been killed by having their necks broken. One specimen (MA2.4) did display signs of a fractured neck; however there was no evidence to show that the adjacent cervical vertebrae had been affected (Fig. 5.4). This fracture must have occurred after death. Consequently, no obvious skeletal evidence to suggest the cause of death could be established in any of the specimens.

5.3 – DOGS

Species Identification

It was not possible to attempt identification of the species for any of the two mummified dogs studied during the course of this project. Mummy MA3.1 appears to have an elongated snout (Fig. 5.5). The cranium is absent in the case of mummy MA3.2.

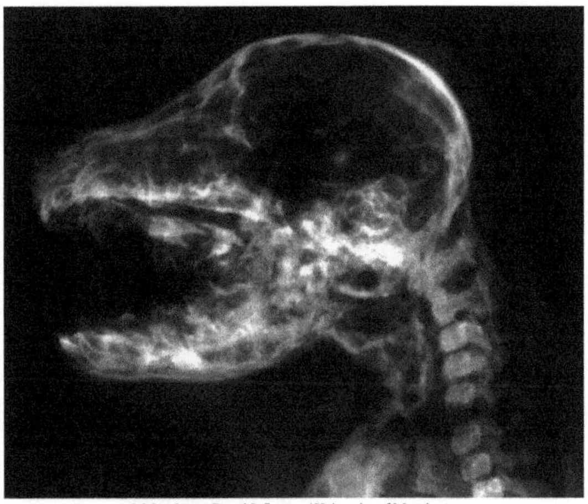

Fig. 5.5 – Lateral radiograph of mummy MA3.1 displaying an elongated snout and concave profile

Ageing and Sexing

One of the dogs (MA3.2) has unfused epiphyses indicating that it is a juvenile animals. MA3.1 presented more of a challenge in terms of the interpretation of the radiographs as the only possible identification of an unfused epiphysis which can be seen on the distal femur, may have been caused by the superimposition of another bone. CT was not able to fully clarify this so it is possible that the animal was juvenile or a very small mature animal.

Body Cavity Contents

No evidence was found of residues in the body cavities; however this is thought to be due to the small skeletal proportions of the two animals studied.

Evidence of Pathological Conditions

The mummified dogs studied revealed one suspected pathological case. The radiographs of mummy MA3.1 appear to show that the teeth were not properly calcified at the time of death which may be an indicator of disease or ill-health (Fig. 5.5). It is unknown as to what medical condition could have caused this phenomena to occur or whether it may simply have been a result of a poor diet low in the necessary calcium to allow the teeth to form normally.

It is also possible that the low density shown by the teeth on the radiographs could have been the result of the treatment received by the animal during the mummification process. Questions arise as to whether it could have been possible that substances applied to the body during mummification somehow caused the decalcification of the dentition. Alternatively, the radiographic process itself could simply be providing misleading evidence for a condition that does not exist. Without dissecting the mummy bundle and removing samples from the teeth for analysis it is impossible to provide a conclusive answer on this subject.

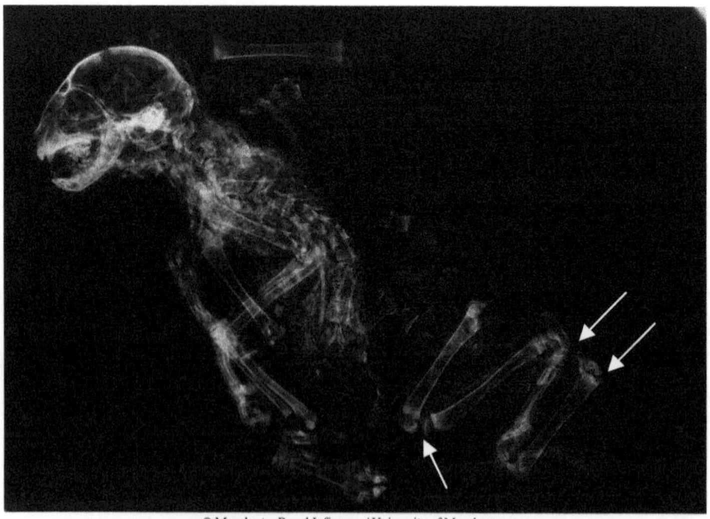

Fig. 5.6 – Lateral radiograph of mummy MA2.22 showing the presence of unfused epiphyses on a number of the limb bones

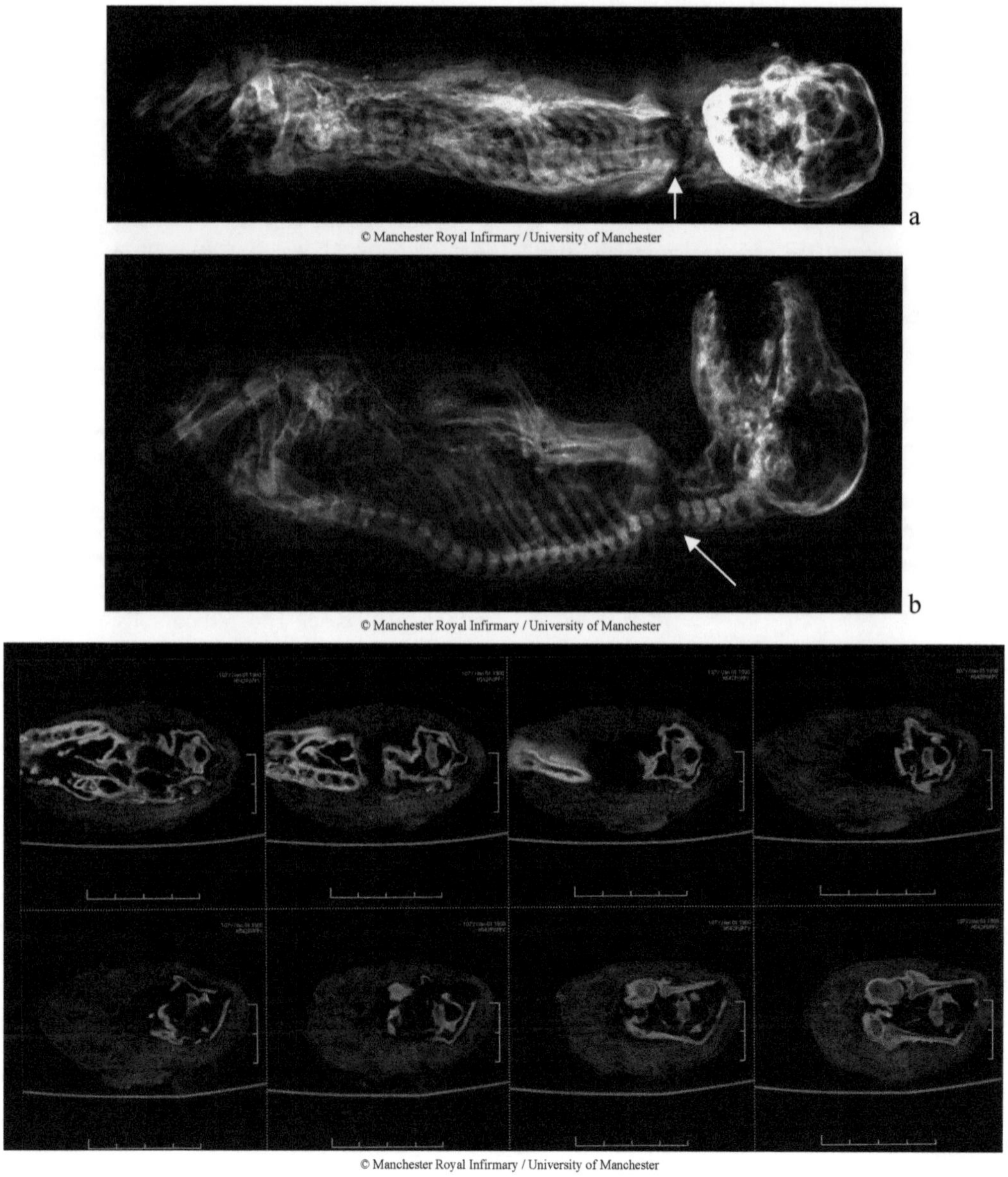

a

b

c

Fig. 5.7 – a) AP radiograph, b) Lateral radiograph and c) CT axial sections of mummy MA3.1 showing the presence of a post-mortem fracture to the cervical spine

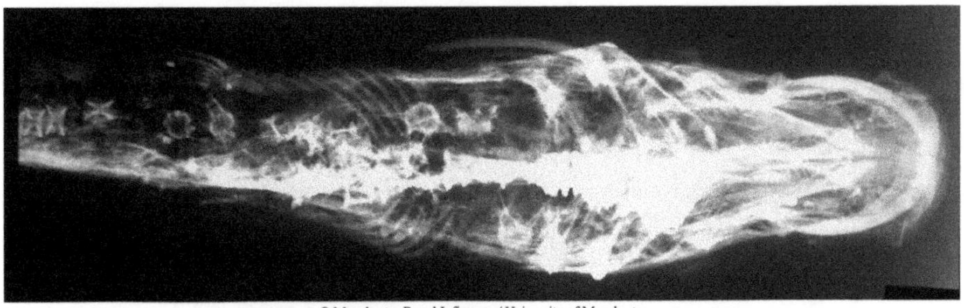

Fig. 5.8 – radiograph of mummy MA4.11 showing the rounded profile of the skull distinctive to the Catfish family (*Clarias)*

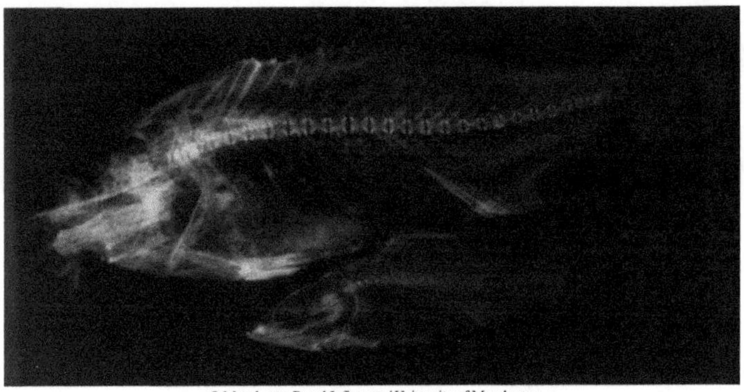

Fig. 5.9 – radiograph of mummy MA4.1 showing the distinctive Nile Perch body shape (*Lates niloticus)*

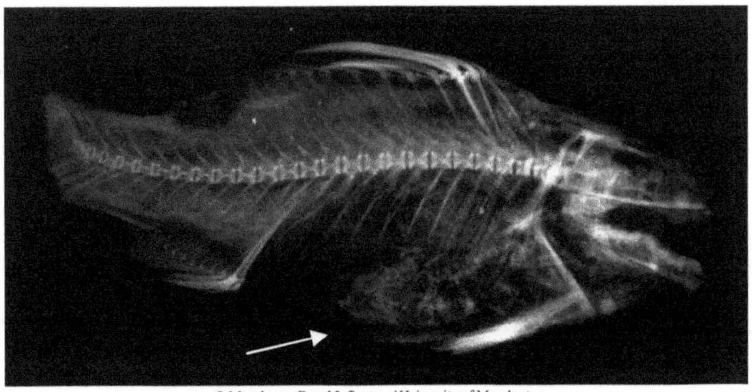

Fig. 5.10 – Radiograph of mummy MA4.5, a Nile Perch (*Lates niloticus*), showing the presence of ingested material in the gut

Evidence of Cause of Death

No evidence was identified to suggest a cause of death for any of the two specimens studied. Mummy MA3.1 does have a fractured neck; however this is almost certainly a post-mortem injury caused by the desiccated condition of the mummy (Fig. 5.7).

5.4 – FISH

Species Identification

Thirteen of the fourteen mummified fish studied have been identified as Nile perch (*Lates niloticus*) which were commonly occurring in the waters of the Nile and could grow to enormous sizes (Fig. 5.9). The exception to this identification are mummies MA4.10 and MA4.11 which are thought to belong to the Catfish family sp. *Clarias* (Fig. 5.8). This fish has a unique appearance compared to the Nile perch as it has a thick and rounded dentary with visible barbells which would have been used to sieve through the silt on the floor of the Nile whilst searching for microscopic foodstuffs.

Ageing and Sexing

It was not possible to attempt ageing of sexing for any of the mummified fish within the study collection.

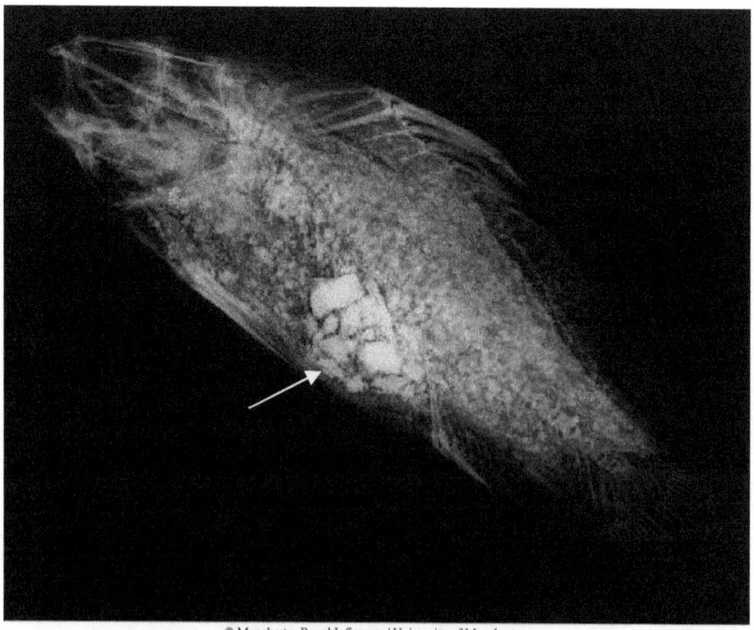

Fig. 5.11 – Radiograph of mummy MA4.12 showing granular masses in the gut area

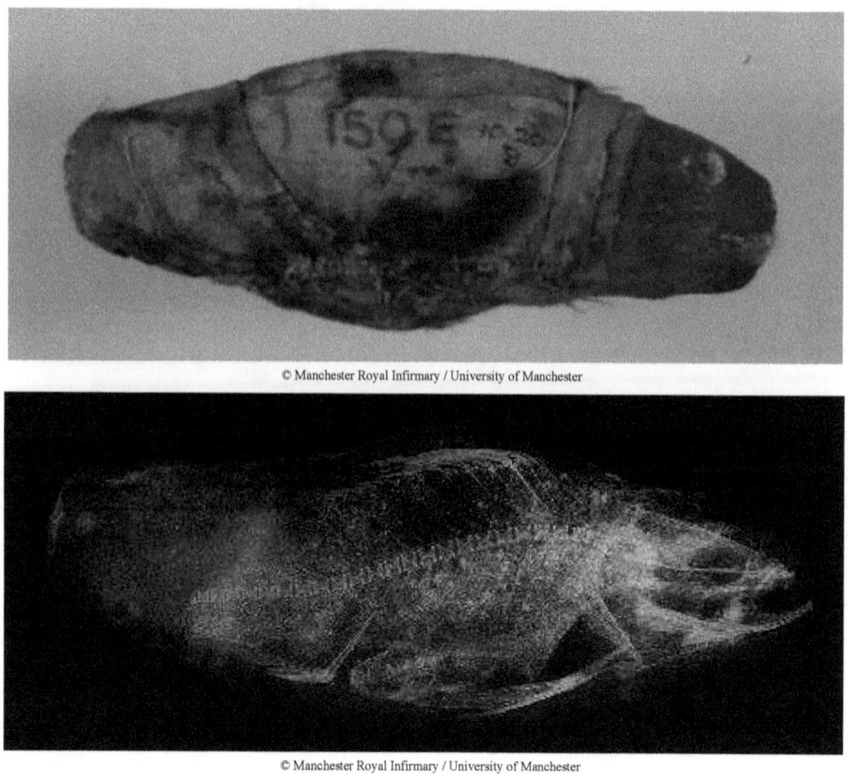

a

b

Fig. 5.12 – a) Photograph and b) Radiograph of mummy MA4.13 which appears to show the presence of a small fish in the gut area. Notice that the embalmers have applied the facial features to the incorrect end of the mummy bundle

Body Cavity Contents

The mummified fish studied provide interesting evidence for the presence of contents in the body cavities. Two of the fish show evidence of ingested material in the stomach area. Mummy MA4.5 shows a mass of small bones and vertebrae indicating that it had consumed another creature not long before death (Fig. 5.10). The stomach of mummy MA4.12 appears to contain a dense granular material thought to be sand or silt from the river bed (Fig. 5.11).

The gut of mummy MA4.13 contains the complete skeleton of a fully-formed small fish (Fig. 5.12). It is thought that the smaller fish must have been inserted into a slit made in the belly, possibly during the evisceration process.

Evidence of Pathological Conditions

No evidence was identified to suggest that any of the fish studied suffered from any pathological conditions.

Evidence of Cause of Death

No evidence was found to suggest the possible cause of death for the fish studied. It is possible that the fish died simply by being removed from the water which would leave no traceable skeletal evidence.

5.5 – CROCODILES

Species Identification

All of the sixteen crocodiles studied are Nile crocodiles (*Crocodylus niloticus*) which were common inhabitants of the waters of the Nile during ancient times.

Ageing and Sexing

Nile crocodiles can grow to as many as six metres in length; however the specimens studied for this project were all juveniles with the exception of mummy MA5.4 which measures 170 centimetres in length and is believed to be a young adult. Many of the smaller crocodiles studied are thought to be hatchlings where others may have been a few weeks or months old.

Sexing was not possible for any of the crocodile mummies studied.

Body Cavity Contents

It was possible to identify a number of examples of body cavity contents within the crocodiles studied. Three mummies (MA5.12, MA5.14 and MA5.15) showed evidence of brain residue in the skull cavity (Fig. 5.13). The radiographic analysis of the young adult crocodile, MA5.4, revealed evidence of a ball-shaped mass in the stomach area which appears to contain dense opacities (possibly stones) and granular fragments (Fig. 5.14). Mummies MA5.6 (Fig. 5.15) and MA5.8 (Fig. 5.16) also show evidence of granular masses in the abdominal cavity. Residue is visible in the orbital cavity of mummy MA5.12 (Fig. 5.17).

Evidence of Pathological Conditions

One crocodile mummy (MA5.10) showed evidence to suggest a possible pathological condition (Fig. 5.18). It is believed that the animal may have suffered from slight

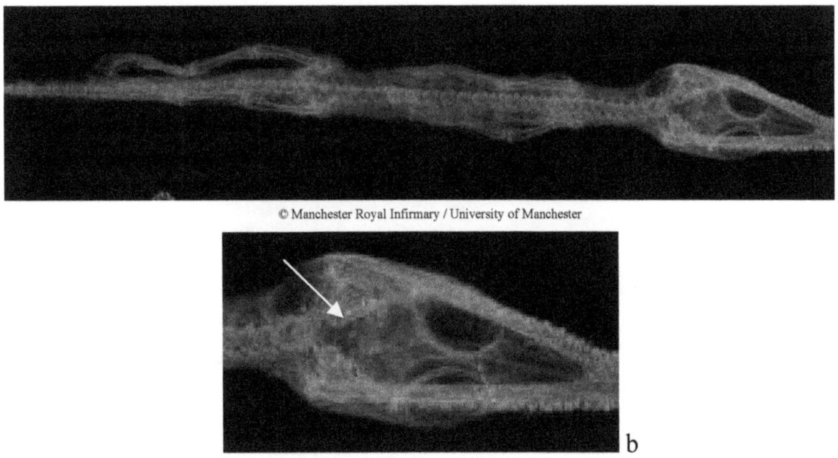

a

b

Fig. 5.13 – a) Radiograph and b) region of interest of crocodile mummy MA5.15 indicating the presence of dehydrated brain matter in the cranial cavity

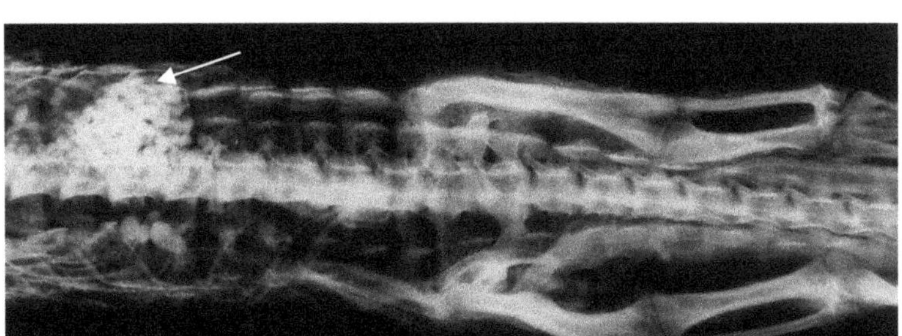

Fig. 5.14 – Radiograph of the mid-section of crocodile mummy MA5.4 showing the presence of ingested material in the stomach

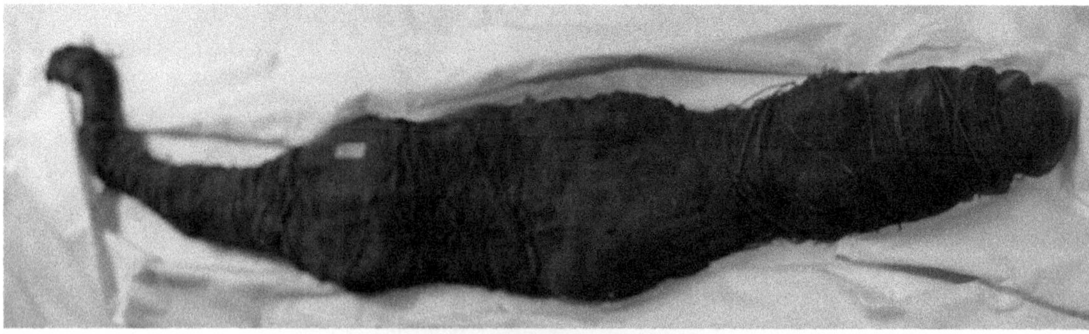

a

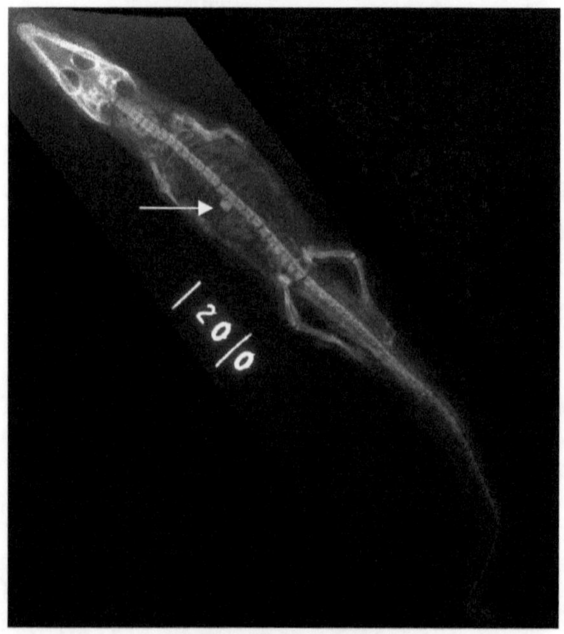

b

Fig. 5.15 – a) Photograph and b) Radiograph of mummy MA5.6 showing the presence of a granular anomaly in the abdomen

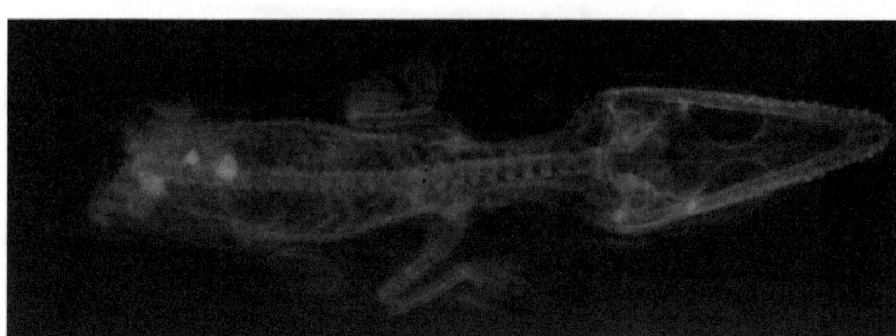

Fig. 5.16 – Radiograph of mummy MA5.8 showing a number of granular inclusions in the abdominal cavity

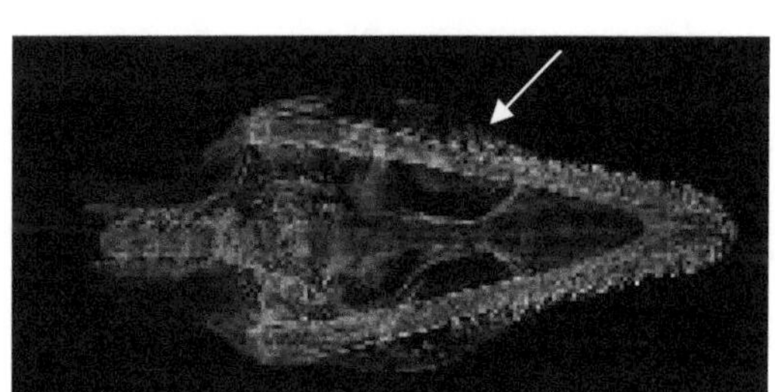

Fig. 5.17 – Radiograph of mummy MA5.12 showing the presence of dehydrated matter in the orbital cavity

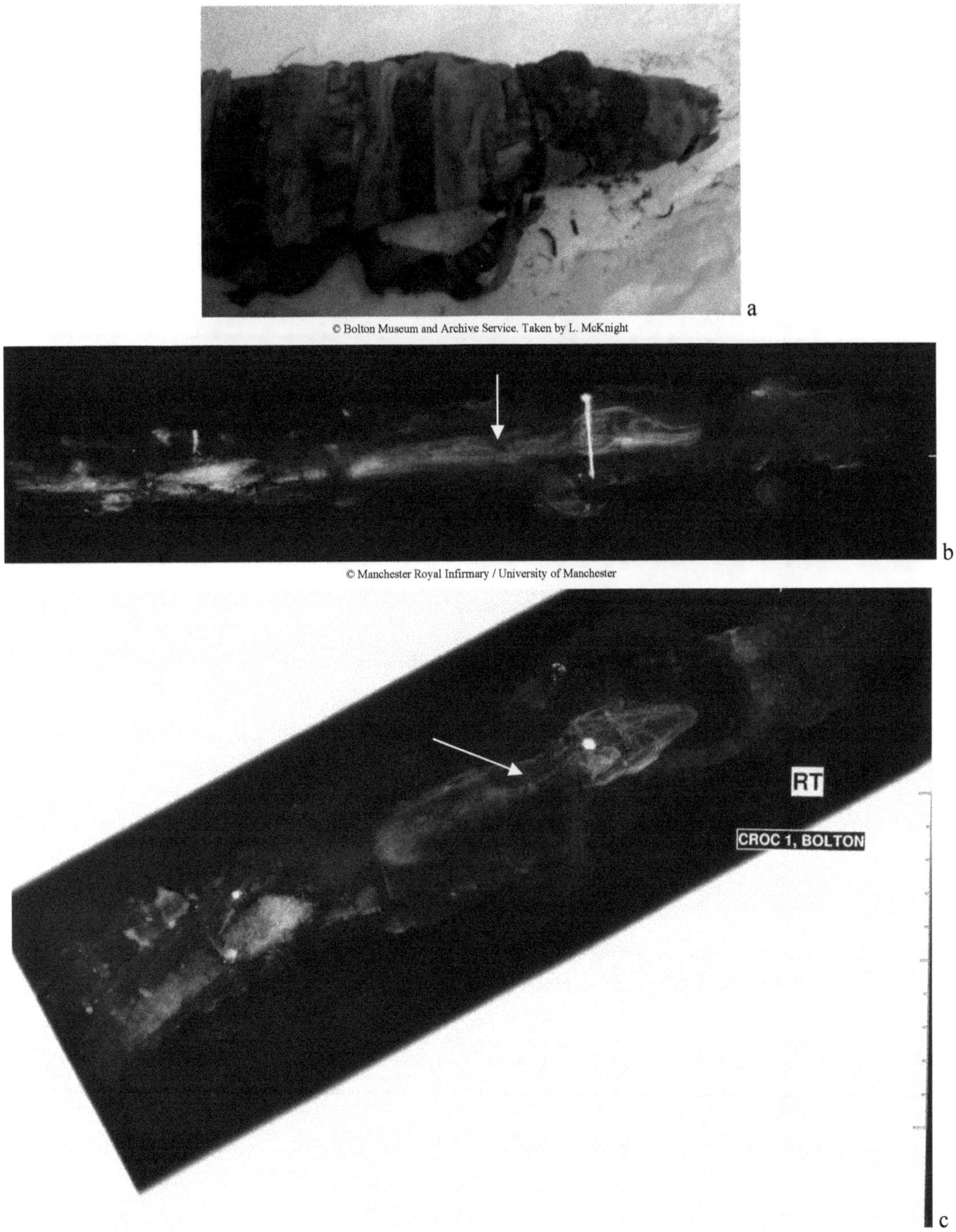

Fig. 5.18 – a) Photograph, b) Lateral radiograph and c) AP radiograph of mummy MA5.10 showing evidence of curvature of the spine. This may represent a pathological condition or it may have occurred as the result of the mummification process

curvature of the spinal column; however the fact that this anomaly may have occurred as a result of the mummification process cannot be completely ruled out.

Evidence of Cause of Death

Crocodiles were greatly feared and it would have been necessary to employ a technique which limited human danger. Herodotus reported that crocodiles were lured onto land using bait and the squeals of a live pig before its eyes were plastered with mud. Once the animal was rendered disorientated and unable to see its attacker, it could be easily dispatched (Selincourt 1985; 156-157; Owen 2001; 103). Methods such as decapitation and blunt-force trauma to the skull would have been the easiest and safest for larger crocodiles.

The young adult crocodile mummy MA5.4 displays extensive fractures to the skull which are believed to be post-mortem; however it is not possible to completely expel the thought that they may represent the cause of death (Fig. 5.19). The radiographs show a linear

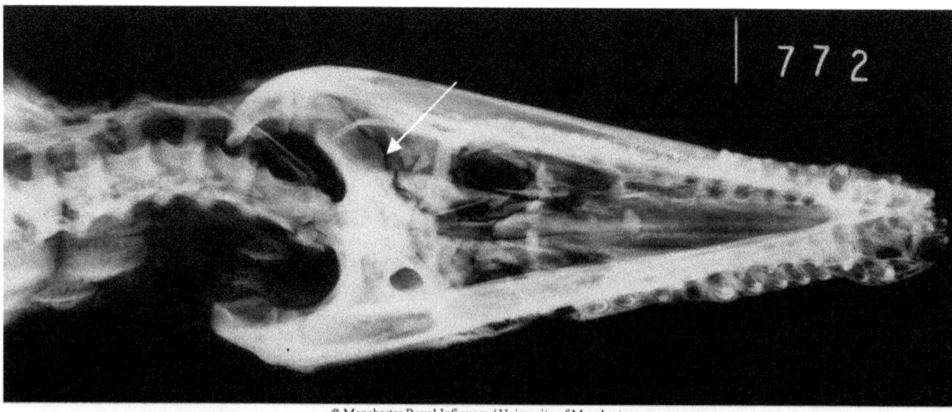

© Manchester Royal Infirmary / University of Manchester

Fig. 5.19 – Radiograph of mummy MA5.4 showing the presence of extensive skeletal trauma to the vault of the skull. This has caused the linear transparency which is visible in the occipital region with disruption to the periorbital bone. It is unclear as to whether this trauma occurred at the point of death or post-mortem

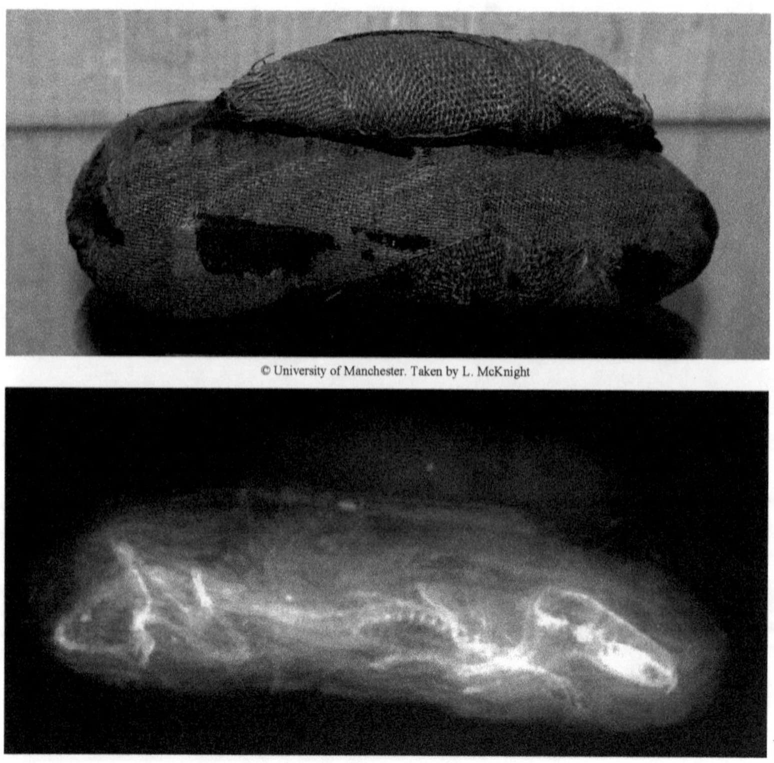

a

© University of Manchester. Taken by L. McKnight

b

© Manchester Royal Infirmary / University of Manchester

Fig. 5.20 – a) Photograph and b) Radiograph of mummy MA6.2 which has been tentatively identified as a shrew (either *Crocidura flavescens* or *Suncus murinus*)

transparency in the occipital region which has caused extensive disruption of the peri-orbital bone. The mandible and maxilla also show evidence of displaced fractures. These types of injuries are consistent with the animal having suffered a blunt-force trauma injury to the skull.

Juvenile animals may have been simply held by the tail and dipped into molten resin or bitumen which seems probable due to the high volume of young animals which display this coating directly on the skin. This would have been a quick and easy method of dispatching large numbers of young crocodiles efficiently whilst in effect mummifying them at the same time.

5.6 – OTHER ANIMALS

Species Identification

The two rodent mummies identified from the radiographs (MA6.2 and MA6.6) were identified by Henry McGhie at the Manchester Museum. MA6.2 is believed to be a shrew (either *Crocidura flavescens* or *Suncus murinus*) (Fig. 5.20) and MA6.6 is believed to belong to the gerbil family (sp. *Gerbilus*) (Fig. 5.21).

No positive species identification was gained for the other mummies.

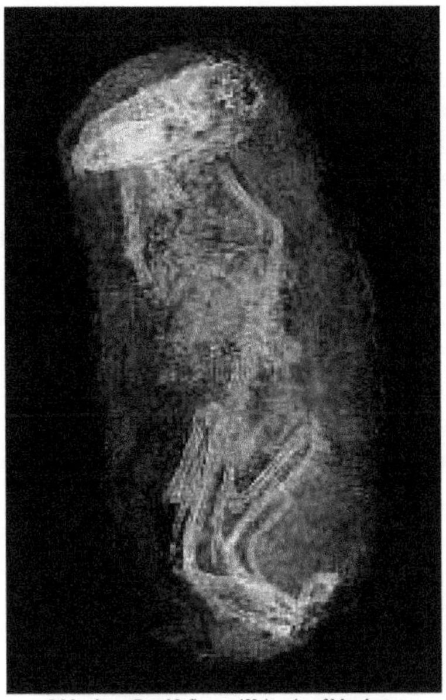

Fig. 5.21 – Radiograph of mummy MA6.6 which has been identified as a member of the gerbil family (sp. *Gerbilus*).

Ageing and Sexing

It was not possible to attempt ageing or sexing on any of the seven animals in this category.

Body Cavity Contents

No evidence was found to suggest contents in the body cavities of any of the mummies.

Evidence of Pathological Conditions

No evidence was found to suggest that any of the animals suffered from a pathological defect.

Evidence of Cause of Death

It was not possible to establish the cause of death for any of the seven animals in this category with the exception of the modern rat mummy MA6.4 (Fig. 5.22). This animal is thought to have died from natural causes before being naturally mummified.

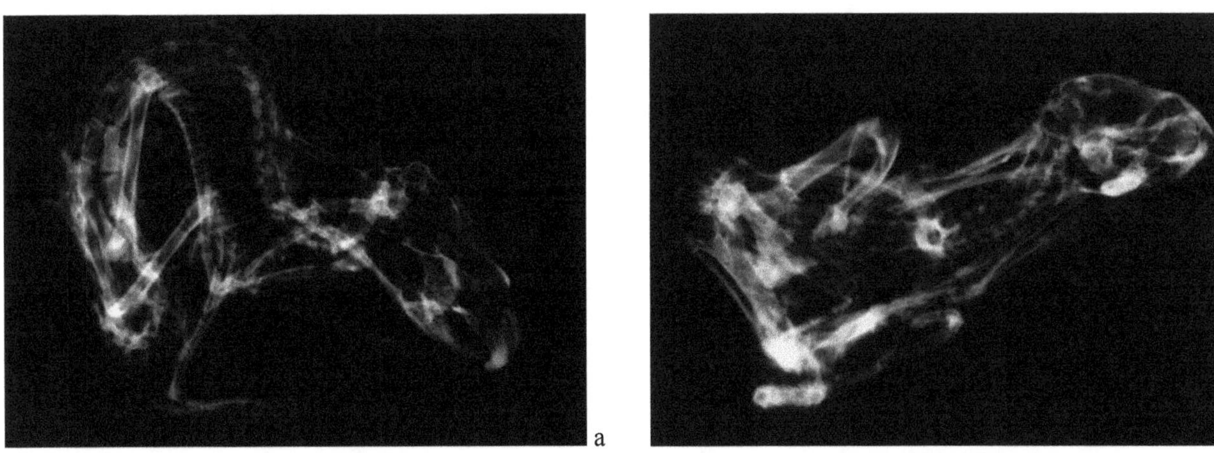

Fig. 5.22 – a) AP and b) Lateral radiograph radiograph of mummy MA6.4 which are the dehydrated remains of a naturally mummified modern rat

5.7 – DISCUSSION

The types of animals identified from the skeletal remains are those which are known to have been worshipped and mummified by the ancient Egyptians. This is not surprising as most previous work done on the mummification of animals has been concerned with the identification of the species themselves. One of the positive points about the study is that there was a broad spectrum of species so it was possible to describe each one in some detail. It was not known at the outset of the project how the species would be distributed and whether the study would be balanced in favour of certain species.

The introduction of specialist knowledge and expertise was vital in helping to establish positive identifications for many of the mummified animals. In cases where the identification has been restricted to sub-species only, this was due to uncertainty over the identifications and it was considered more beneficial to provide a limited identification rather than an incorrect one. It is often impossible to achieve a positive identification using only the radiographs as no direct comparisons can be drawn to reference material. Also, juvenile animals, as many of the study collection represents, can be extremely difficult to identify as the bones have not formed sufficiently (Zivie and Lichtenberg 2005).

Ageing and sexing presented an area of difficulty during this study; however it is not felt that this compromised the research aims in any way. Many animals were classified as juvenile which in itself reveals a great deal about the practice of animal mummification in ancient

Egypt. This discovery also corroborates previous evidence that the animals bred in captivity for ritual purposes were slaughtered before they reached adulthood.

It was not possible to conclusively prove the cause of death for any of the animals. A number of specimens did show evidence of pathological conditions and trauma which could have contributed to their deaths; however it was not possible to suggest whether these were life-threatening or whether they had occurred post-mortem. The majority of fractures which have been noted on the radiographs are likely to have occurred after death and usually following the desiccation of the body during the mummification process.

It was not possible to corroborate theories made by researchers (Armitage and Clutton-Brock 1981) who described how the mummified cats radiographed during their study at the British Museum appear to have had their necks broken.

The healed fracture of the femur shown in the case of mummy MA1.25 is the best example of pathology witnessed in the study collection and represents an interesting case study. No other documented cases have been found where a traumatic fracture of a major bone had been allowed to heal at such an odd angle, leaving the bird with such a major impediment. There is a documented case of an ibis with deformed feet caused by overgrown and twisted claws causing a disability (Leith Adams 1864, 177); however one could surmise that the impediment sustained by MA1.25 was considerably more severe than this.

The presence of contents preserved within the body cavities of a number of the mummies studied suggests that intentional evisceration and removal of the brain was not practised on votive animals. This is thought to be due to the time and effort involved in the procedure which would have slowed down production of the completed mummies. The remaining organs would have been subjected to desiccation through the mummification process with some inevitably being lost to autolysis.

Chapter Six

THE MUMMIFICATION PROCESS

6.0 – THE VARIABLES

Very little literary information regarding the methods of mummification employed on animals in ancient Egypt has survived, so the only resources available to researchers are the animals themselves. Egyptologists are fortunate that such a large number of animal mummies are available for study from a diverse range of archaeological sites. The problem is that the majority of mummies that survive in museum collections have unknown origins and generally dating information is speculative.

The mummified animals available for study are presented in a variety of states of preservation. It is possible for specialist conservation treatments to be used on material of this nature to stabilise further degradation although many of the specimens studied have not received this treatment. Some specimens are deemed too fragile to be transported outside their holding institution for study, and all of those that are transported require careful packing to ensure that further damage is kept to a minimum.

In the early years of Egyptology, dissection was the normal method of analysis for both mummified humans and animals. In terms of ascertaining the method of mummification used, dissection is perhaps the best and most conclusive procedure as it is possible to literally 'see' what the embalmers have done. Obviously, dissection is not recommended as it essentially destroys the mummy bundle, so it is necessary to seek alternative methods of analysis in the form of modern scientific techniques.

There are a number of biomedical techniques available to researchers and they are constantly being redeveloped and refined in the light of new advances and methodological breakthroughs. Radiography remains the only completely non-destructive and non-invasive study method which enables the researcher to visualise the contents of the mummy bundle successfully. The use of X-ray on mummified animals can provide an insight into the method of mummification used, but in the majority of cases, the problems associated with the interpretation of the resulting radiographs outweigh the value of the information they can identify. Evidence to suggest the treatment that an animal received can be difficult to locate on a conventional radiograph or it may not be easily visualised due to the presence of overlying and superimposed features.

Computed tomography (CT) perhaps provides the most promising method for researching the method of mummification used and the ways in which the physical remains were treated. This is because the technique enables the observer to examine localised areas of the body and its wrappings in greater detail and in a variety of different planes without overlap, making interpretation easier. In the case of mummified humans, previous CT attempts have successfully identified the following mummification practices:-

1. The location of the incision through which the internal organs were removed.

2. Highlighted damage caused to the ethmoid and sinus bones during the removal of the brain.

3. The presence or absence of visceral packs within the body cavities which contain the mummified internal organs of the individual.

4. The use of packing materials and symbolic decorative objects placed on or in the body, including items such as false eyes.

5. The technique of wrapping employed by the embalmers.

In the case of mummified animal remains, the general paucity of research carried out to date has meant that very little is known about the process that the remains received during mummification. In the case of large animals, it can be much easier to ascertain how the animal was treated after death simply due to the size of the creature and the associated structures within the body. Being able to identify the presence or absence of the internal organs is obviously much easier in a cow than it is in a cat!

This research project aimed to identify whether radiography has a valid use in ascertaining the method of mummification. It was hypothesised that gaining valuable information relating to certain research questions would be difficult, if not impossible, simply due to the size and condition of the mummies themselves.

External Appearance of the Mummy Bundle

The first method of analysis that should be applied to all mummified animal material is a thorough assessment of the external appearance of the mummy bundle. This can help to suggest the species of the animal that it is intended to represent. In the majority of cases, the external appearance of the mummy forms the basis of the museum records, so any species identification attempted at this stage is purely by observation. Radiography provides a non-destructive method to confirm the species identification speculated on the basis of the external assessment, and in some cases it shows that the mummy contains either a partial skeleton or no skeletal material whatsoever.

Mummy Shape

The study of ancient Egyptian mummified animals has identified certain body positions which appear to have been considered 'standard' by the embalmers. These standard positions for the display of the physical remains seem to have been designed to provide the most easily recognisable body form whilst being relatively simple to construct using the skeletal form as a framework. This is a logical explanation for the existence of standardised body positions in animal mummies when one considers that the pilgrims visiting the temple complexes with the intention of buying an offering to the deity would have preferred a physically identifiable object. The pilgrims were essentially ordinary people with firmly held beliefs and an animal mummy which outwardly resembled the animal itself would undoubtedly have strengthened their attachment to the votive object.

Different species of animal were mummified in different ways by the embalmers. Obviously, the major factor which determined the shape of the finished mummy was the physical form of the animal's body. For example, the long flat body of a crocodile lends itself perfectly to a slim and elongated mummy bundle, whereas a cat naturally forms a plumper, more cylindrical bundle when wrapped. Essentially it is possible to say that the embalmers were working with the materials they had, yet at the same time they moulded what seem to have been socially accepted standardised animal body forms.

In some cases, the shape of the mummy was ultimately determined by the container that it was manufactured to fit inside. One notable example of this practice are the falcon mummies from cemeteries such as Saqqara which were stacked from floor to ceiling in cone-shaped ceramic pots. The cylindrical form of the mummy bundles themselves fit snugly into the pots and would have been easily produced and replicated by the embalmers.

There are exceptions to these standardised forms where the skeleton of the animal is positioned in a different way to create a completely different mummy bundle shape. Often these are examples where the animal is positioned in a more anatomically natural position, yet they do not fit the embalmers' normal template so they appear to be different.

During the course of this research project, it was discovered that those mummies which were intended to realistically portray the form of the animal itself were those that, when analysed using radiography, were found to contain no skeletal material. There were a number of examples of this within the study collection which will be discussed below. Some of them appeared to be so stylised in their external appearance that it became possible to identify them as pseudo-mummies prior to them being imaged. The following sub-sections are intended to provide some information on the research questions which were being addressed. These are followed by sections which relate to each species of animal individually in terms of how they were treated during the mummification process.

Position of the Skeleton

The position in which the skeleton lies within a mummy bundle appears to have followed certain standardised practices in the majority of cases. It is hypothesised that this was largely due to the fact that votive mummies were produced hurriedly to meet a high demand and as such it became necessary to adopt methods which enabled this process to work efficiently. The use of standardised 'templates' would have achieved this.

Obviously, the position of the skeleton is largely influenced by its shape and size and generally relates to how the limbs, wings or head were positioned in relation to the torso. The techniques employed on each species will be discussed below.

Wrapping

Animal mummies were usually wrapped in some form of linen bandaging. It is not uncommon to find animal mummies which have been treated (and sometimes wrapped) before being placed directly into a holding vessel such as a ceramic jar or stone-cut niche. On a practical level, wrapping provided a means of sealing the mummy bundle, making it more appealing for sale purposes and it also meant that it was possible to decorate the exterior of the mummy bundle.

Researchers who study mummified animal material know all too well that their subjects are presented for study in all sorts of conditions and states of preservation. In many cases it is possible to see bones protruding from within

thin layers of bandages, whilst other specimens appear to be tightly wrapped. Some have only a loose covering of linen, where others have formed rigid packages where the bandages are used to provide support to the mummy bundle itself.

The most elaborate methods of wrapping tend to be witnessed during the Roman period of Egyptian history when the votive mummy industry was reaching the peak of its popularity (Ikram and Iskander 2002, v). Many of the mummies studied in this research project are believed to date from the Roman and Ptolemaic periods.

Evisceration

Evisceration, or the removal of the internal organs, was a major stage of the mummification process, yet one which was not employed at all time periods. Human mummification can be classified into chronological periods depending on whether or not evisceration was carried out. Animal mummies are a little more obscure with the vast majority not being eviscerated at all. This is likely to be because the largest group of mummies, the votive animals, were the product of hasty production methods put in place to ensure that there was maximum output in the minimum amount of time. Evisceration was probably considered by the embalmers to be an 'unnecessary luxury' for which there seemed to be little need. These were, after all, offerings sold in their thousands to temple pilgrims. Also on a practical note, it would not have been possible to eviscerate many of the small animals.

The cult animals worshipped by the ancient Egyptians were treated with lavish mummification methods similar to those used on humans. These animals were extraordinary symbols of ancient Egyptian worship, occupying positions of social status similar to the ruling pharaoh. It is hardly surprising that, upon death, they were treated to the time-consuming mummification techniques that their human counterparts also received.

It is highly probable that, in existing cases where there is evidence to suggest that the internal organs were removed, they were not kept for any ritual purpose. This may explain why canopic jars used to hold the viscera have only been identified for the sacred animals and not for any other category of animal mummies. The practice of the removal of the brain is also difficult to prove as, although many of the larger sacred animals show empty crania, no reasons have been identified to suggest how this was achieved. Normally this would have been achieved by breaking through the ethmoid bone and allowing the brain to be pulled out of the cranial vault through the nostrils; however, alternative methods in larger animals may have involved dislodging the atlas vertebra and removing tissue via the foramen magnum (Ikram and Iskander 2002; iv).

It can be extremely difficult to assess whether an animal has been eviscerated through the use of radiography. Radiographs can be notoriously difficult to interpret because of the problems of superimposition and the overlapping of features, and limited contrast resolution can make evisceration very difficult to identify. When this is coupled with the fact that the mummification process inherently alters the composition of the internal organs, irrefutable evidence for evisceration becomes increasingly difficult to confirm. Dehydrated organs assume a similar radiographic attenuation to bone, skin and muscle following the mummification process, making it difficult to distinguish between them. The larger the animal being studied, the more likely it would be to attempt to identify evisceration, simply because the physical components within the body are likely to be more easily visualised. The majority of mummified animals studied during the course of this project were small.

CT offers the best opportunity of identifying evidence of evisceration in mummified animals and humans alike. This is because it enables the body to be viewed in axial sections enabling critical areas suspected to have been treated to be scrutinised in more detail and in a variety of different planes which can make interpretation easier. In human mummies, it is possible using CT technology to identify the incision made by the embalmer to enable the retrieval of the internal organs. This may be possible in the case of large animals, although I have found no documented cases where it has been attempted, let alone been successful.

In this research project, the best hope of finding evidence for evisceration lies in the identification of areas of the body where cavities have been filled with packing materials during the mummification process. This packing, providing that it can be demonstrated that the material has been inserted into a cavity rather than being applied to the exterior of the body, indicates that an incision must have been made by the embalmers. In cases such as these, it is plausible to conclude that evisceration took place to some degree, with packing being used to bulk out the body to form a natural shape.

Packing of the Body Cavities

Evidence for packing of the body cavities is important as it demonstrates that an element of care and attention was lavished upon the animals during the mummification process. Little is known about the packing materials used by the embalmers except those uncovered during early autopsies of mummified remains. Packing of the body cavities is likely to have been used to accentuate the natural form ensuring that the body kept a recognisable shape following the mummification process. The incision made for the purpose of packing the body cavities may indicate that evisceration was practised.

Radiography can be helpful in identifying the types of materials used to pack the body cavities. It is common to find evidence for grit or stone particles within the packing substance. It is likely that the substance consisted of mud or sand which would have been readily available to the embalmers.

Inclusions within the Mummy Bundle

Researchers studying ancient Egyptian mummified animals understand that these objects are often found to contain surprising items. Chapter Seven will discuss the existence of so-called 'pseudo-mummies' in more detail, but firstly it is important to mention the mummies which contain surprising inclusions. Inclusions are items which do not necessarily 'belong' in the mummy bundle, but that have been placed there intentionally or have been included by accident. These inclusions could be anything including small stones, amulets, or miscellaneous bones.

Inclusions found within the mummy bundle could be either intentional or unintentional. In the case of items such as amulets or bones, it seems reasonable to suggest that they were placed there intentionally for purposes unknown. Smaller unidentifiable anomalies such as stones or grit particles are more likely to have become embedded within the mummy bundle accidentally during the mummification process. Unintentionally placed objects tell researchers little about the mummy itself, but can highlight practices followed by the embalmers. For example, animals which show evidence of having been dipped in sticky bituminous substances following death and prior to mummification, often show the presence of granular material which has been attracted to the sticky layer.

The results of the radiographic analysis are given below. Separate sub-chapters have been assigned to each of the different species identified to enable the reader to select relevant information more easily and in the hope of providing a more comprehensive review of the radiographic findings.

Fig. 6.1 – Photograph of mummy MA1.21 depicting Thoth seated on a throne surmounted by the Atef-crown

6.1 – BIRDS

External Appearance

There appears to be evidence of a broad spectrum of wrapping techniques and decorative methods used on the birds within the study collection. This ranges from a thin layer of bandages to which a tar-like substance has been applied, to very elaborate geometric designs covering the length of the mummy bundle. These geometric designs are often constructed from bandages in different colours to make them appear more decorative and appealing to the pilgrims.

In a number of cases, an appliqué pattern made from fabric has been applied to the exterior of the wrapped mummy bundle. The patterns are of imagery of the gods to which the animals were held sacred. One of the ibis mummies from the collection (MA1.21) displays such appliqué decoration which is thought to represent Thoth seated on a throne surmounted by the Atef-crown (Fig. 6.1). Little is known about this practice, but one can only assume that it was intended as a means of strengthening the connection between the mummy and the deity.

There are a number of examples of false feet and heads being given to the bird mummies to create the illusion that the animal within is larger and to make the exterior of the bundle appear to be more lifelike. MA1.1, a mummified falcon thought to date to the Late Period, displays a false head and feet which appear to have been treated with a form of gilding or resin, highlighted by a thin radio-dense layer on the radiographs (Fig. 6.2). MA1.19 contains the partial and fragmented remains of a kestrel which has been given a carefully modelled false head. MA1.20, the mummy of an ibis, has an elaborate false head which had been realistically modelled with a long slender neck and beak (Fig. 6.3). An Atef-crown is displayed behind the false head.

Mummy Shape

The shapes given to the mummified birds studied appear to conform to the same pattern across the study group. The small birds of prey, falcons and hawks, are all wrapped in a cylindrical shape with the tail feathers creating the length of the bundle. This body shape has been documented at the large animal necropolises where bird mummies were found deposited in conical ceramic

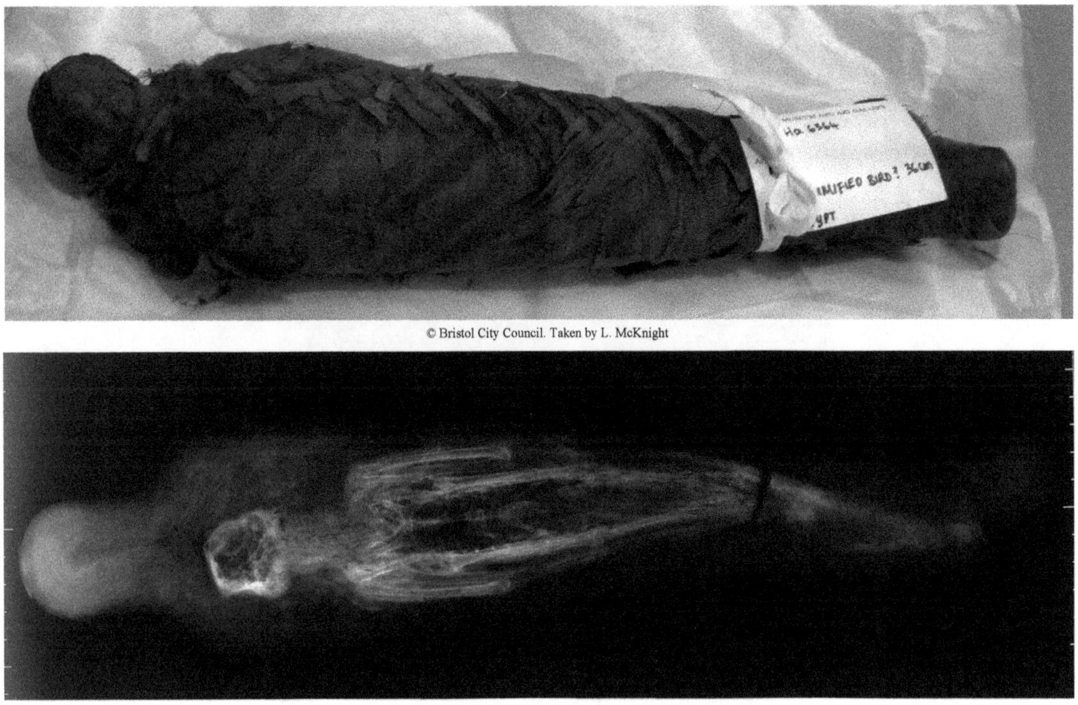

a

© Bristol City Council. Taken by L. McKnight

b

© Manchester Royal Infirmary / University of Manchester

Fig. 6.2 – a) Photograph and b) AP radiograph of mummy MA1.1 showing the false head and feet

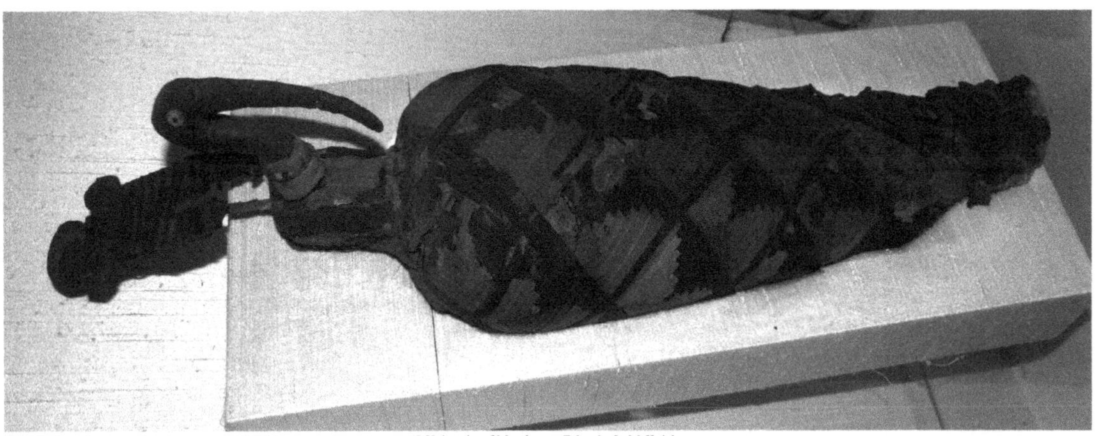

© University of Manchester. Taken by L. McKnight

Fig. 6.3 – Photograph of mummy MA1.20 clearly showing the elaborate false head and the Atef-crown behind

jars. The mummified ibises seem to have been given a broad and flat shape which reflects the larger body size and long neck and limbs of the bird in life.

A number of the birds have been given extra layers of bandages to provide additional support to their fragile skeletons and to provide rigidity to the completed mummy bundle. Falcon mummy MA1.1 has been provided with a neatly constructed roll of fabric to each side of the neck demonstrating care and attention on the part of the embalmer.

Position of the Skeleton

Generally, the birds studied appear to have been mummified with the legs stretched down, the wings folded in close to the torso and the head in a natural position as in the case of MA1.22 (Fig. 6.4). The cylindrical shape formed would have enabled the bird mummy to have been placed conveniently into a ceramic pot of the kind found at the large necropolises such as Saqqara.

A couple of exceptions to this form were identified during the radiographic process. Mummy MA1.4 has the legs flexed up in front of the torso which gives the bird a squat appearance despite the fact that the mummy bundle itself is long enough to have accommodated the animal with the legs down (Fig. 6.5). A similar skeletal position is identified in the case of MA1.8 where the legs have been flexed to such an extent that the claws appear on the images to be close to the head (Fig. 6.6). In this case the head appears to have been forced down on itself which has caused displacement and gives the impression of

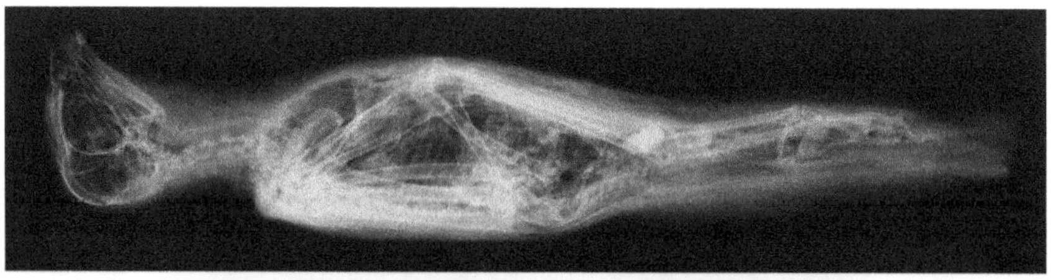

a

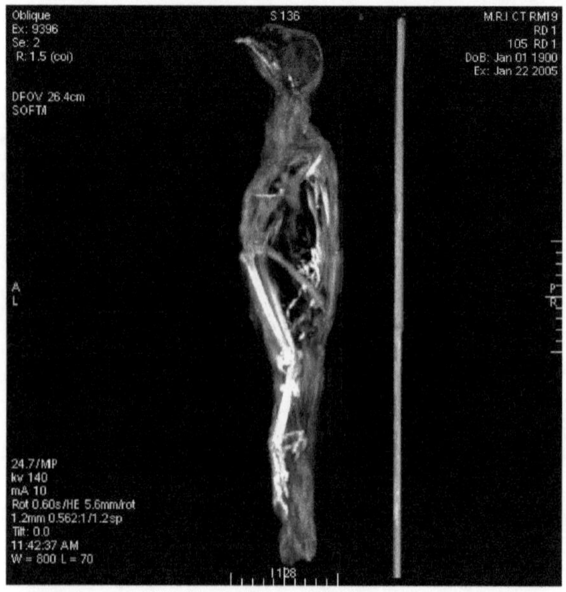

b

Fig. 6.4 – a) Lateral radiograph and b) CT reconstruction of mummy MA1.22 demonstrating the tightly wrapped cylindrical form often given to mummified birds. Image b) depicts the bones and remaining soft tissue on the bird

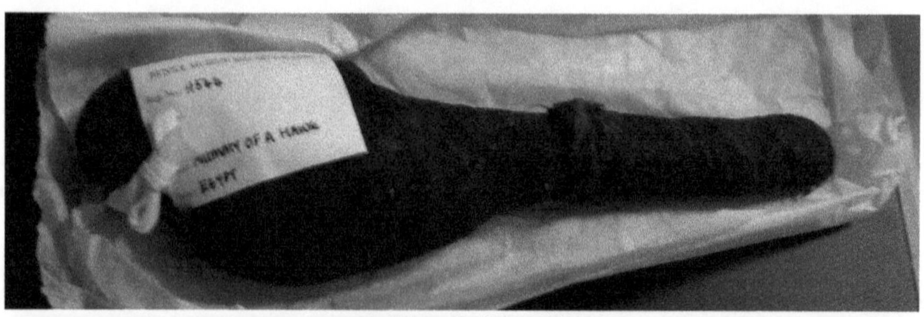

a

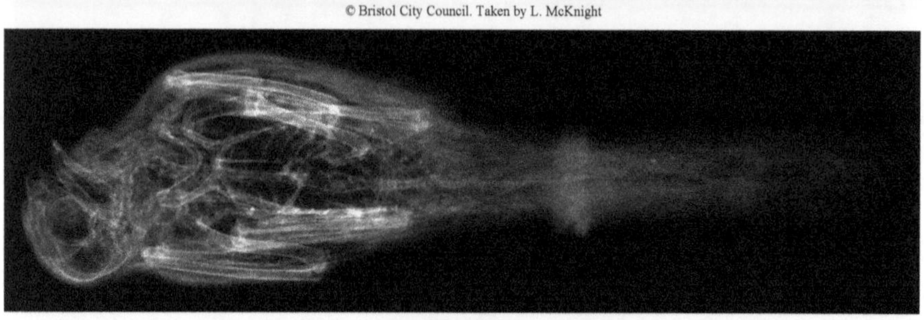

b

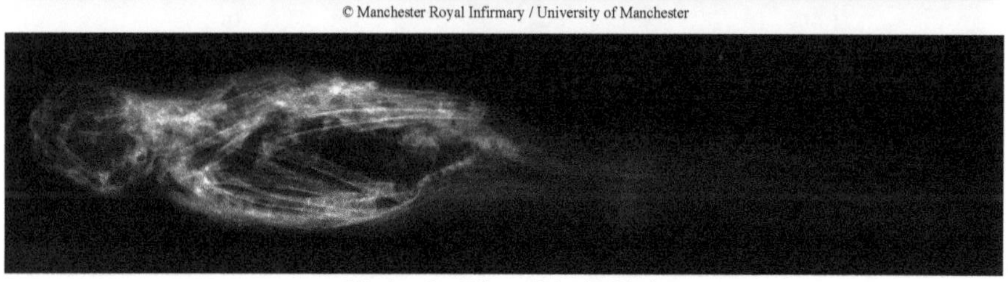

c

Fig. 6.5 – a) Photograph, b) AP radiograph and c) Lat radiograph of mummy MA1.4 showing the bulbous body form of the mummy bundle caused by the underlying skeleton having its legs flexed back on themselves

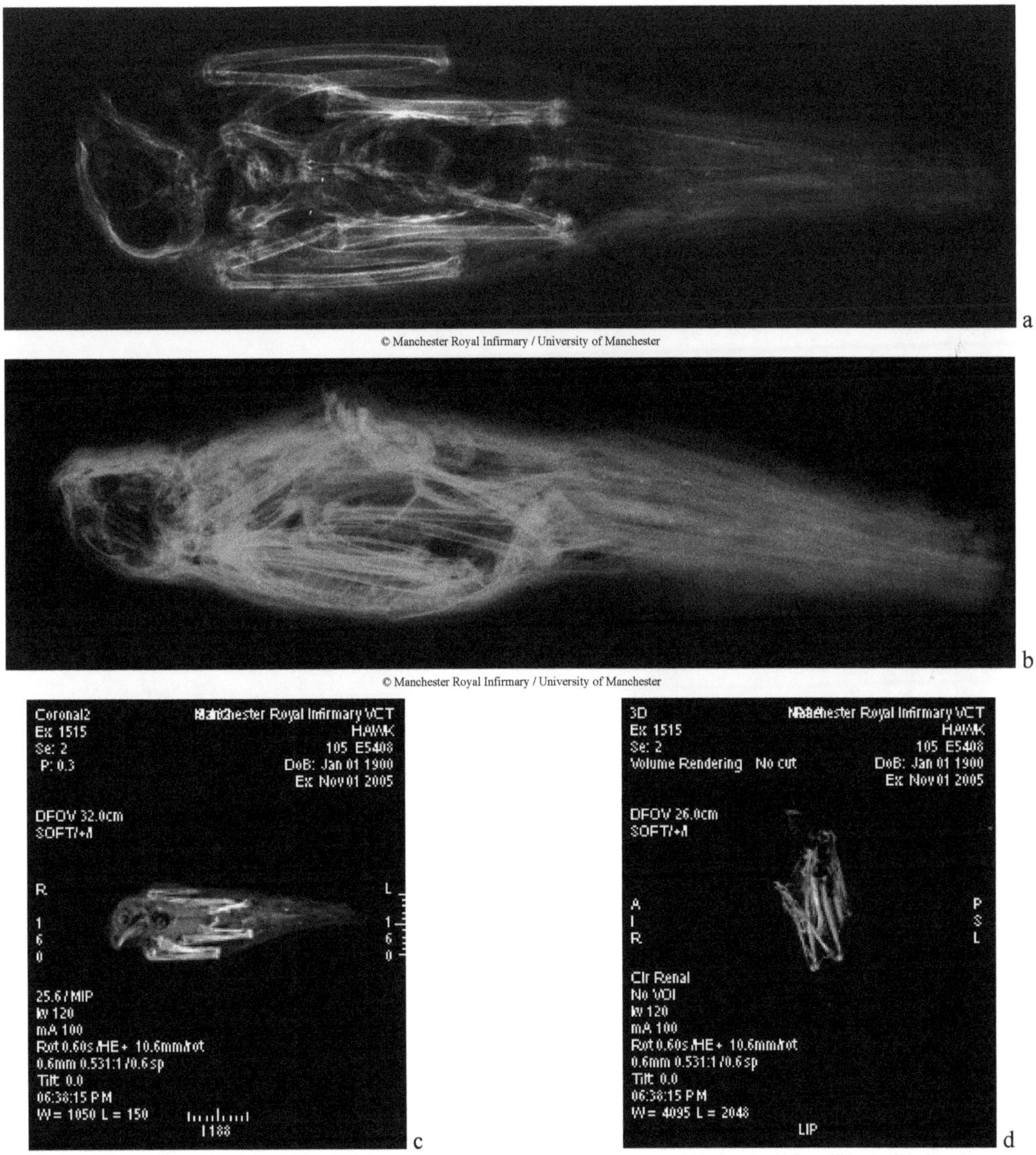

Fig. 6.6 – a) AP radiograph, b) Lat radiograph, c) Coronal reformation and d) volume rendering image of mummy MA1.8 showing the legs flexed back on themselves with the tail feathers being used to give length to the mummy bundle. Image d) gives a 3-D insight into the contents of the bundle

fore-shortening. The limbs are also flexed in an unnatural position in MA1.14 (Fig. 6.7).

The only complete and largely articulated skeleton of a Sacred Ibis (*Threskiornis aethiopica*) that was identified during the study (MA1.20) appears to have had the skull detached from the torso and placed between the legs. This contradicts Leith Adam's description of the ibises he studied in the mid-nineteenth century which commonly had the head brought down and placed on the breast of the bird (Leith Adams 1864; 179). The reasons behind this decapitation are unclear; however, it can be speculated that the removal of the skull and its subsequent position between the legs would have saved space by making a more compact mummy bundle.

In the case of mummy MA1.23 the skull appears to have been folded closely into the chest rather than being allowed to stand naturally (Fig. 6.8).

Wrapping

The bird mummies within the study collection revealed evidence of a wide range of bandaging techniques. The majority were tightly wrapped; however, in a number of cases the beak had become exposed.

One of the bird mummies studied (MA1.1) appears to have been wrapped in stages. This practice can be evidenced on the radiographs through the identification of clearly defined layers often varying in the degree of

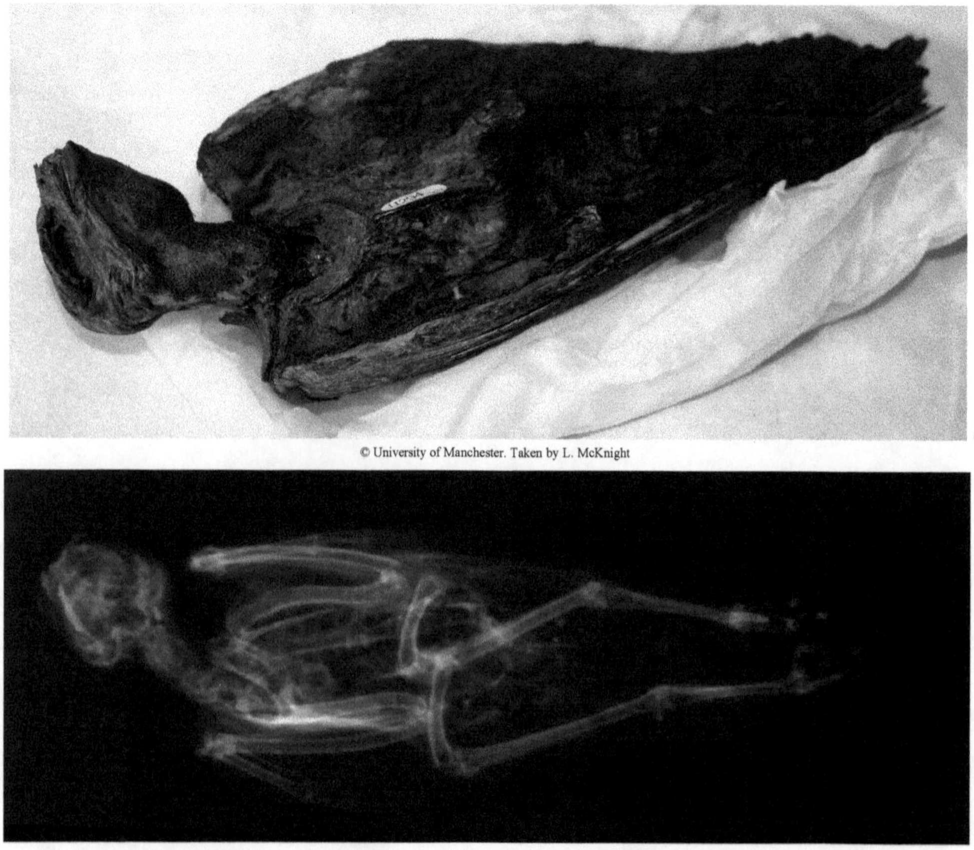

Fig. 6.7 – a) Photograph and b) AP radiograph of mummy MA1.14 demonstrating an unnatural body position with the legs bent and twisted to the side

compression applied. The images show that there are three clearly defined layers of bandaging. The falcon was wrapped initially in a very thin layer of fabric, possibly a single layer, which appears to have been treated with a radio-dense substance such as natron, paint or tar. This bundle has then been wrapped again to form a more cylindrical shape with a roll of bandages being added to each side of the neck for support. The bundle was then wrapped for a third time and a false head and feet have been added.

The use of feathers, although not strictly speaking a form of wrapping, has been identified on the radiographs by the presence of trapped air within the quills. They appear to have been used to provide additional support to the fragile skeletal remains and to add length to the finished mummy bundle. The presence of feathers appears to be better demonstrated using CT technology than X-ray, as in the case of mummy MA1.35 (Fig. 6.9).

Evisceration

It was not possible to identify any evidence for the practice of evisceration from the bird mummies in the study collection. It is suspected that none of the bird mummies had been eviscerated due to the hurried nature of their dispatch and mummification as opposed to the techniques employed in their study not yielding positive results.

Packing of the Body Cavities

No evidence was found to suggest that the body cavities of the birds had been intentionally packed by the embalmers. In the cases where the body cavities do show evidence of contents, it is considered to be dehydrated matter which was present in the cavities at the time of death.

Inclusions

A number of interesting inclusions were highlighted through the radiographic process of the bird mummies.

The mummy bundle belonging to one of the falcons in the collection (MA1.28) was found to contain the distal half of a crocodile skeleton which appears to have been intentionally placed along the breastbone of the bird during the mummification process (Fig. 6.10). Interestingly, the radiographs of this mummy revealed that the bird may have suffered from a defect to the spinal column. It is possible that the embalmers witnessed this defect and were attempting to provide additional support to the bird's weakened posture. This mummy in particular provided an excellent case study through which to demonstrate the potential value of radiography and especially CT. Computer software was used to alter the attenuation of the images to gain a clearer insight into the different skeletal components represented within the bundle.

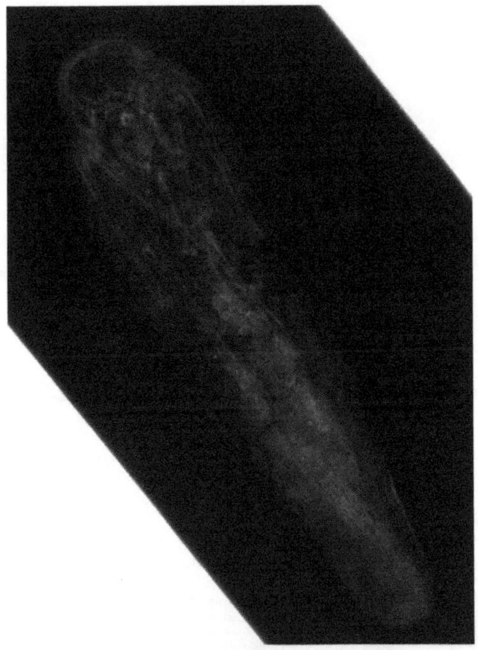

a

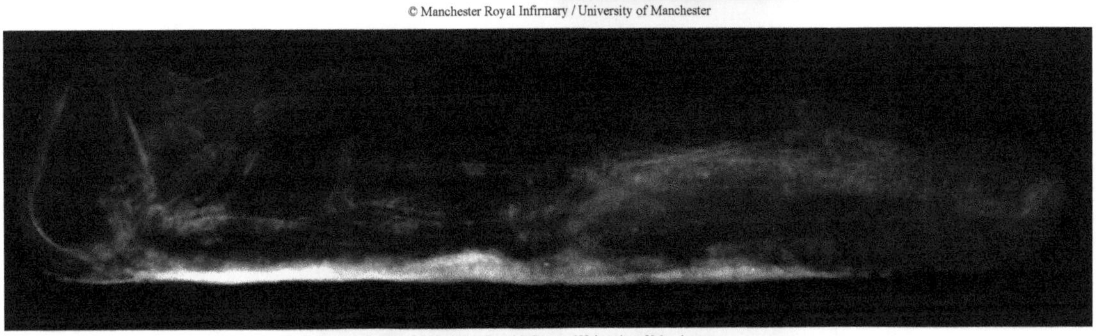

b

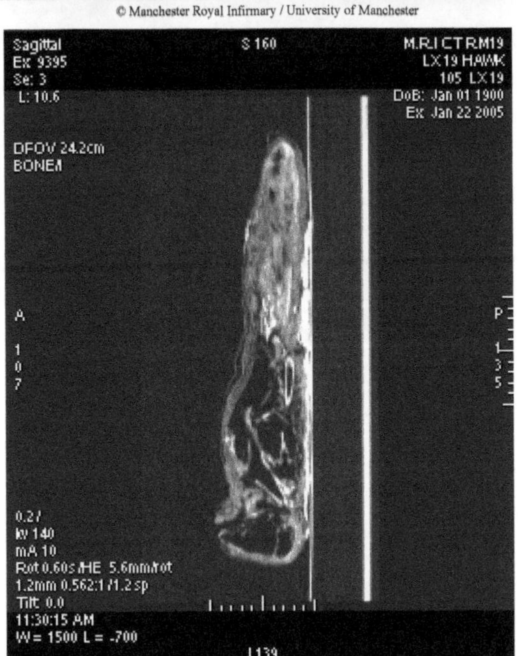

c

Fig. 6.8 – a) AP radiograph, b) Lat radiograph and c) Sagittal CT reformation of mummy MA1.23 showing the skull forced down onto the breastbone

Mummy bundle MA1.2 was found to contain the decapitated skeleton of a bird wrapped alongside a mass of smaller bones which are thought to be the skeletal remains of an insectivore (Fig. 6.11). The skull of the insectivore in particular can be clearly seen on the radiographic images, but few of the remaining small

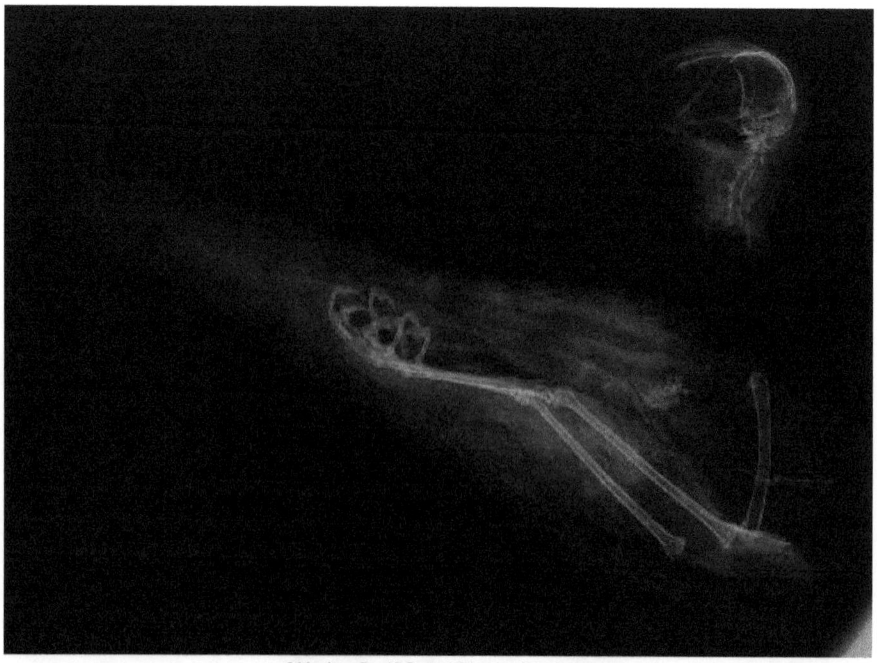

Fig. 6.9 – Radiograph of mummy MA1.35 showing how the pinion feathers have been used to create the length of the mummy bundle in the absence of the complete skeleton

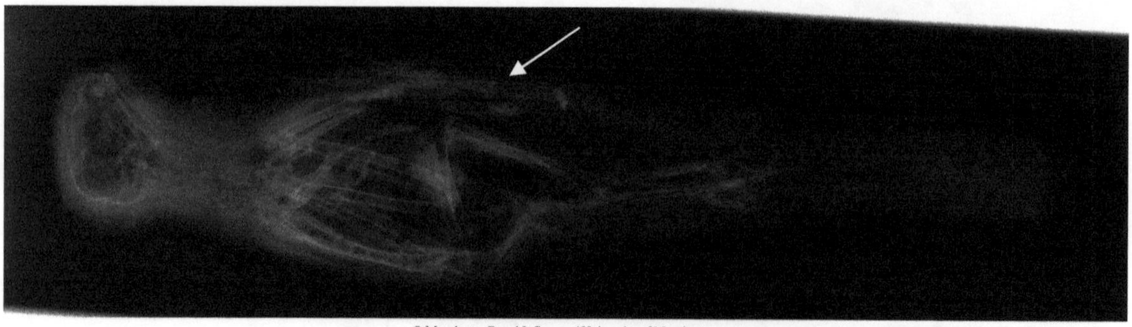

Fig. 6.10 – Radiograph of mummy MA1.28 showing the inclusion of the distal half of a crocodile skeleton along the bird's breastbone

bones are identifiable, although it is thought that they represent the jumbled remains of the remainder of the animal.

Feathers appear to have been used in many cases to reinforce and strengthen the shape of the mummy bundle. Mummy MA1.35 was found to contain only the partial remains of a falcon skeleton yet pinion feathers had been added (possibly from a different animal altogether) to provide the necessary shape. It is often possible to identify feathers from radiographs as the air that gets trapped both in the shaft and between the quills appears as linear areas of opacity. A similar practice has been used on MA1.7 which contains the jumbled and incomplete remains of two birds wrapped together as one with pinion feathers being used to create length and provide support.

The radiographic examination of MA1.1, the falcon mummy with a false head and feet, showed interesting anomalies (Fig. 6.12). The embalmers appear to have wrapped the bird in three distinct layers of bandaging, with the outermost layer including a row of inclusions. These inclusions have almost certainly been placed carefully in a line in an intentional act for reasons unknown. CT revealed that these inclusions are very thin and it is thought that they could be small wax discs.

6.2 – CATS

External Appearance

In total, eight of the fourteen complete mummy bundles found to contain feline skeletal material had been decorated elaborately to make them aesthetically pleasing. In a number of cases the embalmers have obviously afforded a great deal of time and effort in making the exterior of the mummy bundle as attractive as possible to potential purchasers.

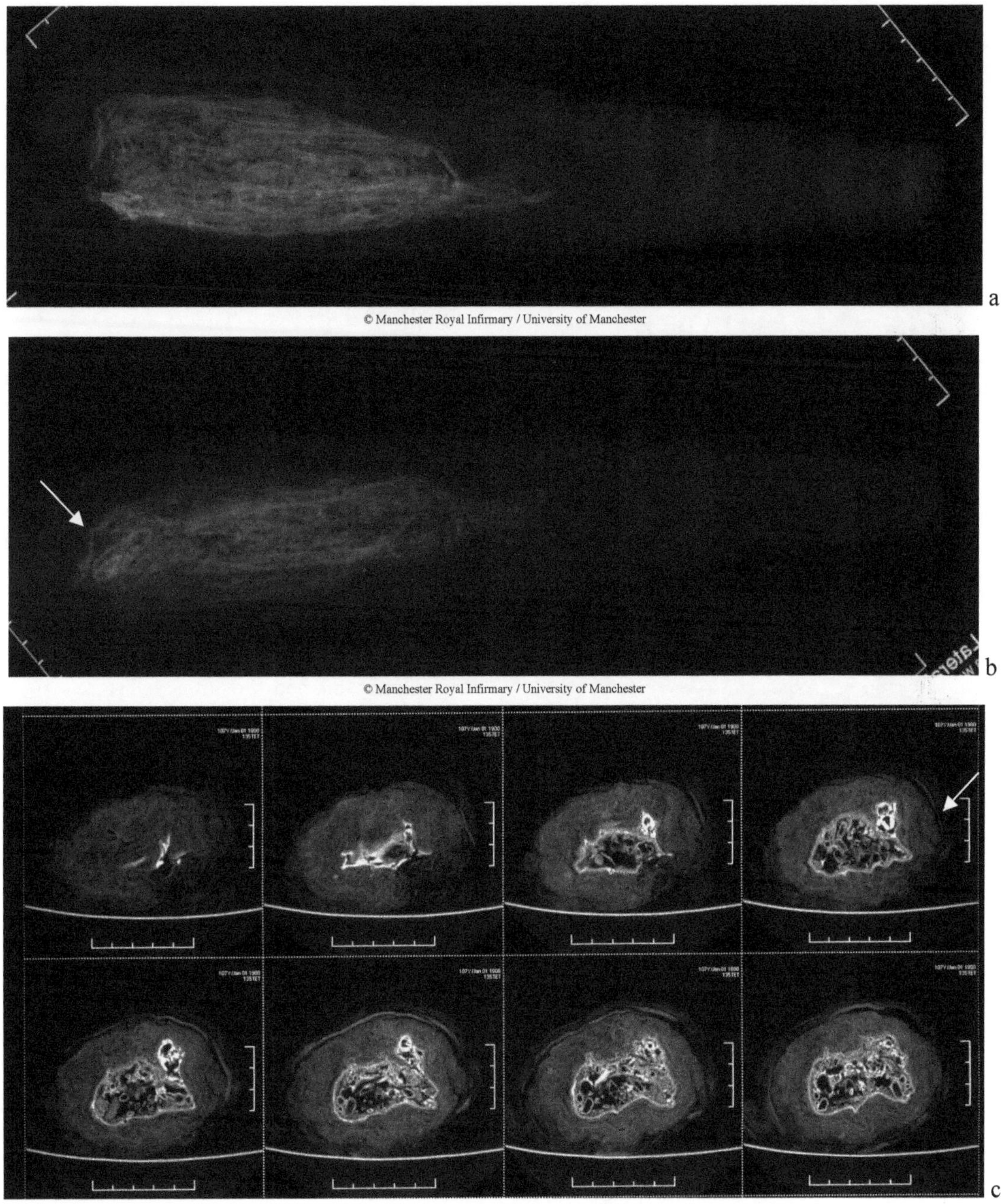

Fig. 6.11 – a) AP radiograph, b) Lat radiograph and c) CT axial sections of MA1.2 showing the decapitated bird skeleton with the partial skeleton of an insectivore wrapped together. The arrow on image c) shows the skull of the insectivore

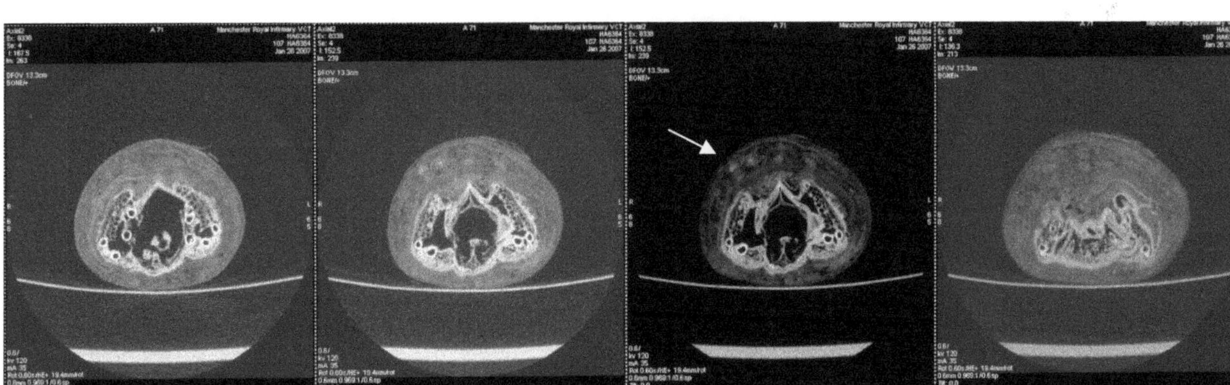

Fig. 6.12 – CT of mummy MA1.1 showing a row of inclusions which are thought to consist of a wax-like substance

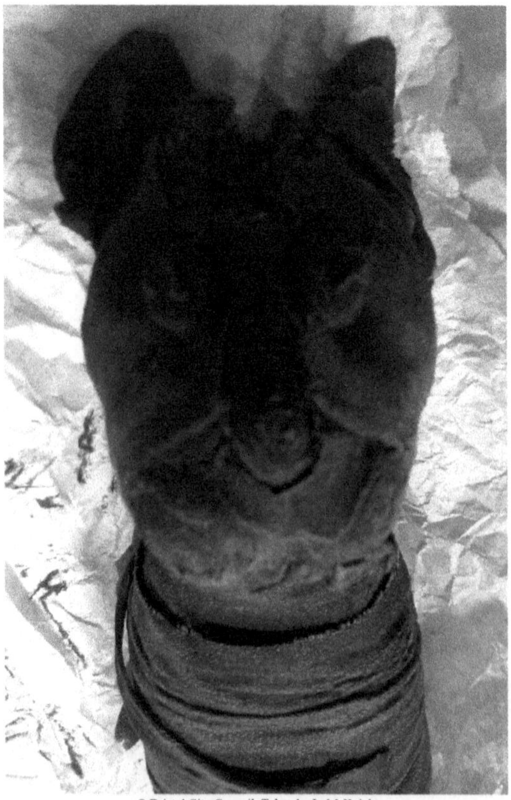

Fig. 6.13 – Photograph of mummy MA2.3 showing the elaborate detail given to the facial area with the addition of eyes, ears and a nose modelled from linen

A number of the mummies studied have been given facial features added to the exterior of the wrappings. These can take the form of painted features, usually black or red in colour, which appear to be stylised in nature. Alternatively, the facial features appear to have been modelled from linen before being fixed down, giving the face a three-dimensional appearance. Four of the cat mummies studied show ears modelled from linen giving them a more realistic appearance.

Interestingly, MA2.3, which is the roughly bandaged mummy of a cat, has been given an extraordinary amount of detail to the facial area (Fig. 6.13). This attention to detail effectively hides the fact that the bundle contains only a partial and disarticulated skeleton. The embalmers may have been attempting to over-compensate for the substandard physical remains interred within by creating a beautifully modelled face with cloth ears and linen buttons for eyes.

Mummy Shape

The cat mummies studied have all been wrapped in a cylindrical form with the obvious exception of disarticulated skulls and the unwrapped kitten (MA2.13).

Position of the Skeleton

Cats appear to generally be positioned with the hind limbs bent upwards and the front limbs folded straight down in front of the torso. The tail is generally folded up between the lower limbs and to the front of the chest (Armitage and Clutton-Brock 1981). Mummy MA2.12 demonstrates this body position well (Fig. 6.14).

Interestingly, in the case of MA2.18, the tail is not present on the radiographs and it is believed that it must have been removed post-mortem due to the resulting bone margins (Fig. 6.15). The reasons for this action are unclear.

Wrapping

A number of the cat mummies studied appear to have been wrapped in stages in a similar manner to that witnessed with the bird mummies. Radiography highlighted that mummy MA2.5 has been wrapped twice with a layer of a radio-dense substance (possibly bitumen, resin or tar) evident between the two layers of linen (Fig. 6.16). A similar double wrapping technique is shown on mummy MA2.1 although in this case there is no evidence

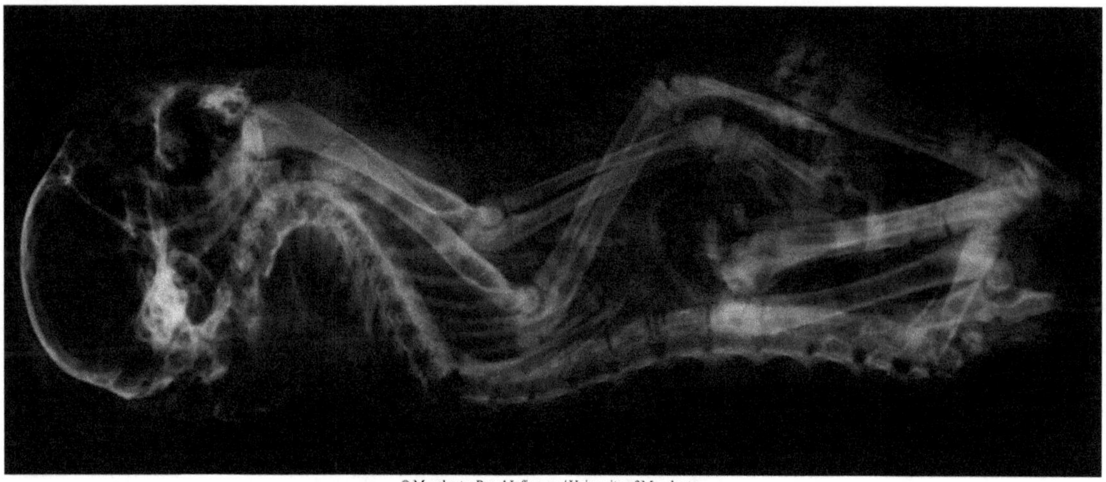

Fig. 6.14 – Radiograph of mummy MA2.12 showing the familiar position of the skeleton often witnessed in feline mummies

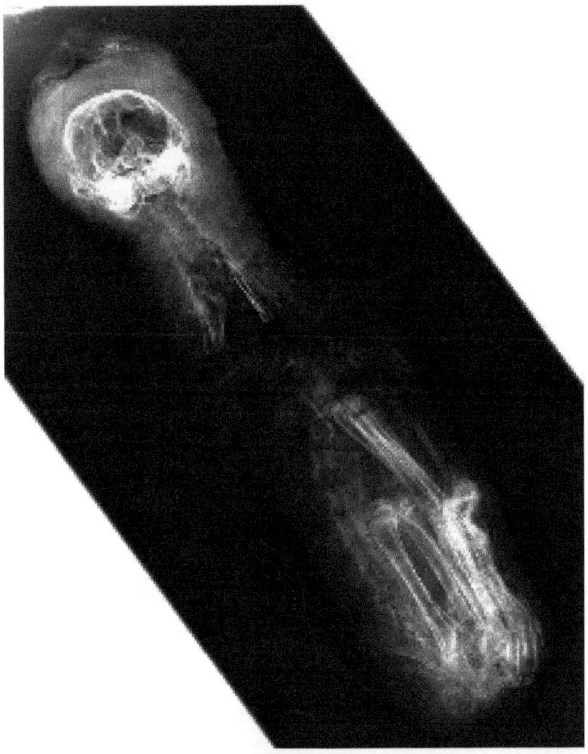

a

© Manchester Royal Infirmary / University of Manchester

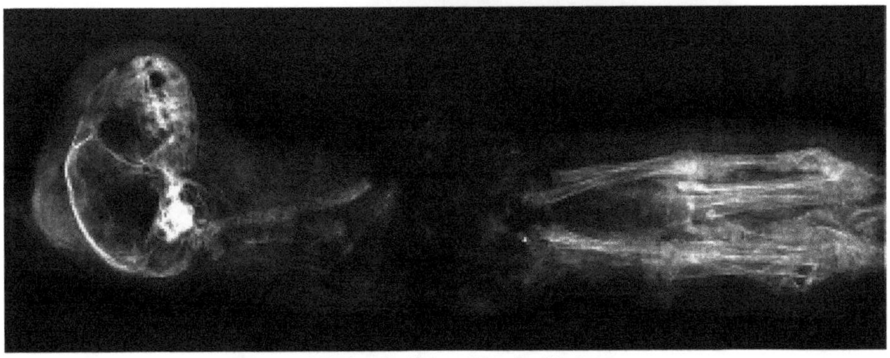

b

© Manchester Royal Infirmary / University of Manchester

Fig. 6.15 – a) AP radiograph and b) Lat radiograph of mummy MA2.18 showing the absence of a tail within the mummy bundle

to suggest that the bundle was treated between the two layers of wrappings. Three distinct layers of bandaging are visible on mummy MA2.2 which was highlighted during the analysis of the CT images (Fig. 6.17).

The CT analysis of MA2.3 clearly shows that rolls of fabric have been used to provide the body shape in the absence of a complete skeleton (Fig. 6.18).

Undoubtedly the most intriguing cat mummy studied during this project is that of MA2.6, also known as the Chester Mummy (Fig. 6.19). This is one of a limited number of mummies studied for which documented records existed although these related more to the acquisition of the mummy rather than to its original provenance. The plainly wrapped mummy bundle is covered with an elaborate net of lapis lazuli beads. The presentation of the mummy in an anthropoid wooden coffin immediately set it apart from all the other mummies studied and this, coupled with its large size, had caused earlier researchers to believe that it contained the remains of a small child. This theory was proved to be incorrect in 1973 which it was studied using radiography at the University of Liverpool and was found to contain the remains of a large cat (Information from Museum archives). The mummy is currently on display at the Garstang Museum, University of Liverpool. MA2.6 will be discussed in greater detail in Chapter Nine.

Evisceration

No evidence was found in any of the twenty-one mummified cats studied to suggest that evisceration had been practised.

Packing of the Body Cavities

One cat mummy (MA2.15) provided evidence that the body cavity may have been packed prior to wrapping; however, it is impossible to assess whether the area of

a

© Bristol City Council. Taken by L. McKnight

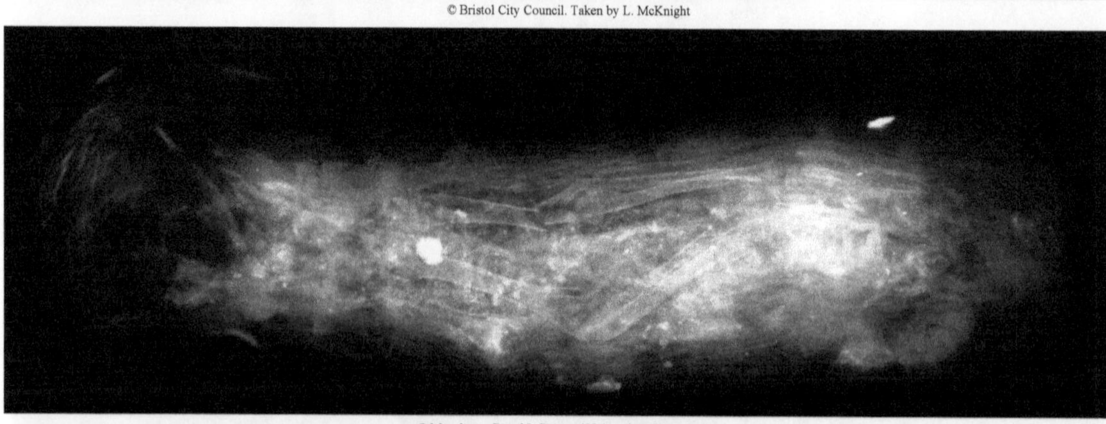

b

© Manchester Royal Infirmary / University of Manchester

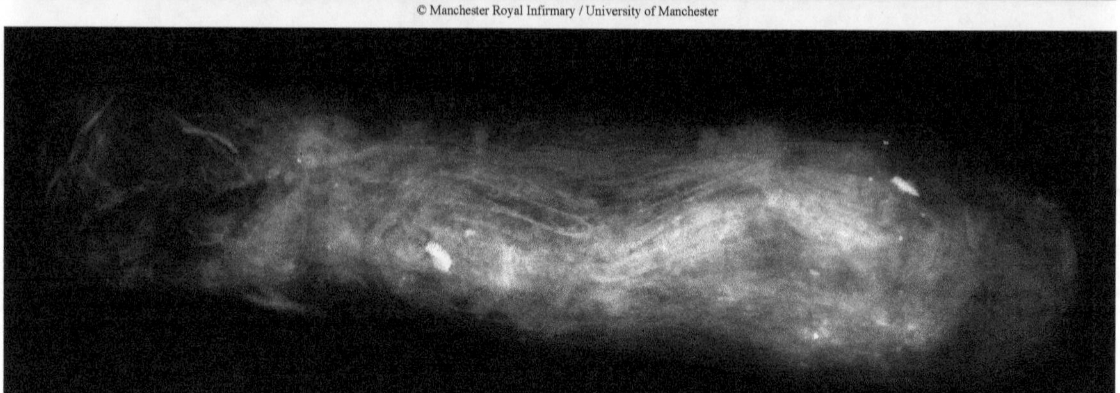

c

© Manchester Royal Infirmary / University of Manchester

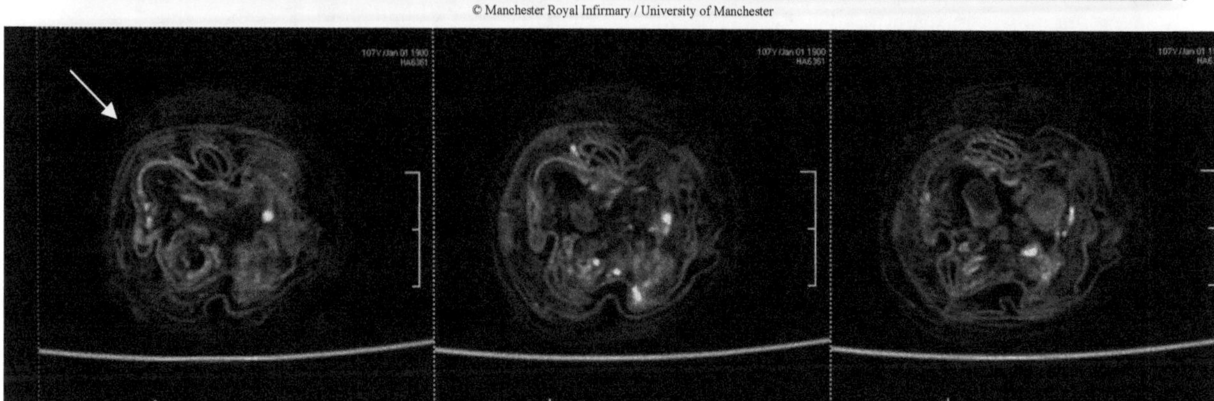

d

© Manchester Royal Infirmary / University of Manchester

Fig. 6.16 – a) Photograph, b) AP radiograph, c) Lat radiograph and d) CT axial slice images of mummy MA2.5 showing the multiple layers of bandages used in the preparation of the mummy and the apparent use of a radio-dense substance. Image d) enables the viewer to gain a clear insight into the bandaging techniques employed

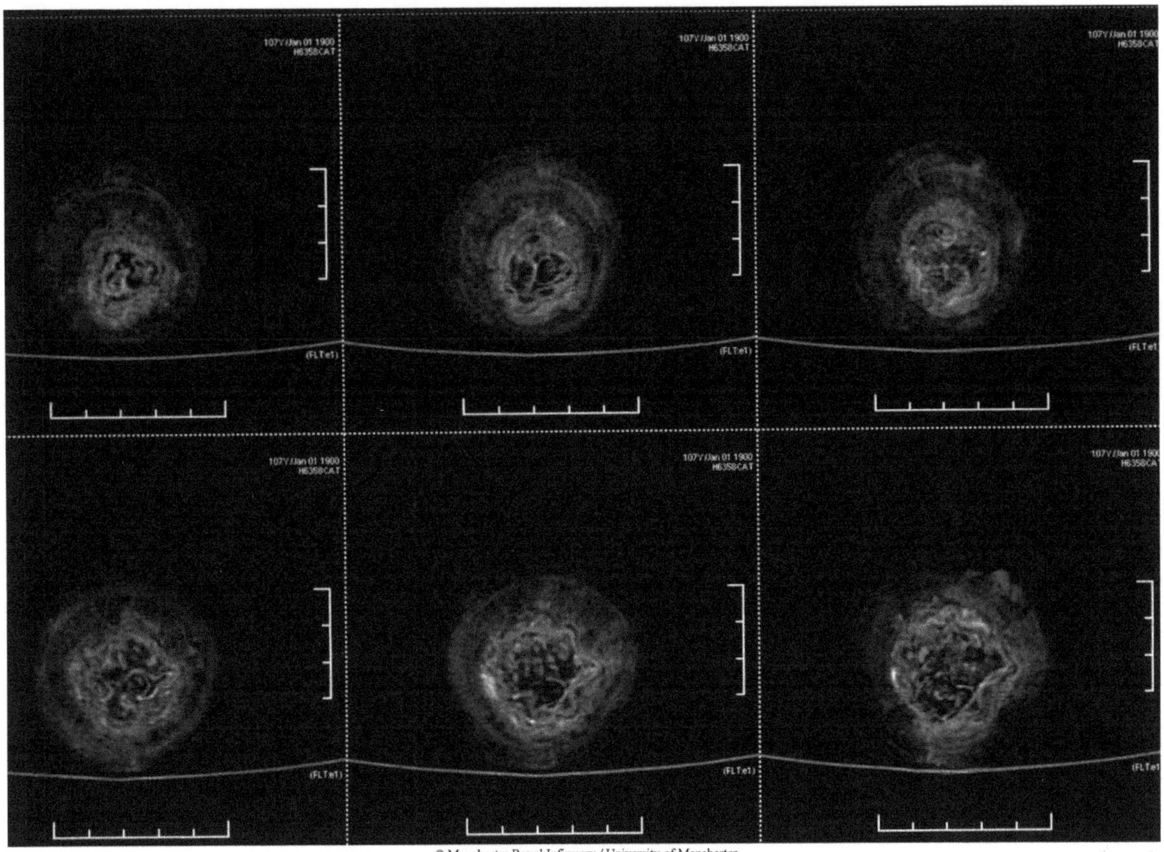

Fig. 6.17 – CT axial slice images of mummy MA2.2 showing the three distinct layers of bandaging used to wrap the bundle

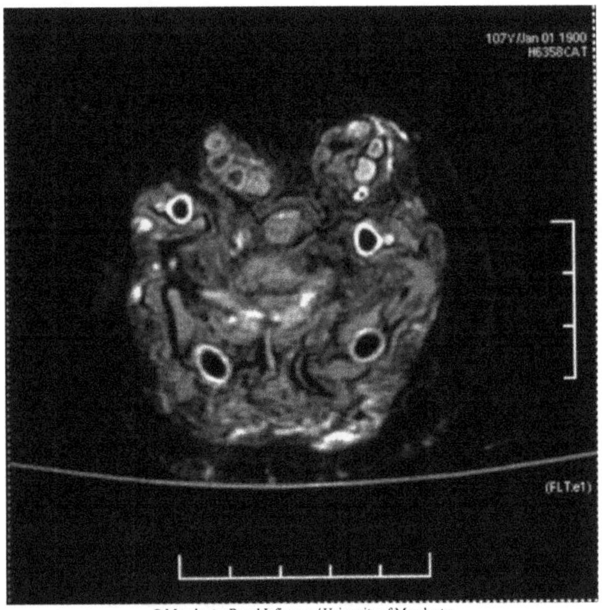

Fig. 6.18 – Axial slice image of mummy MA2.3 showing the extensive use of fabric within the bundle to create the body form in the absence of the complete skeleton

opacity witnessed on the radiographs can be attributed to this activity. It may represent an area of resin which has solidified during the dehydration process. Detailed analysis of the radiographs and the CT images has been unable to clarify the nature of this anomaly.

Mummy MA2.8 contained the skeleton of a juvenile cat which demonstrates the presence of small pellets in the throat area (Fig. 6.20). It is likely that these pellets represent an unidentifiable material ingested by the animal shortly prior to death. It is not possible to

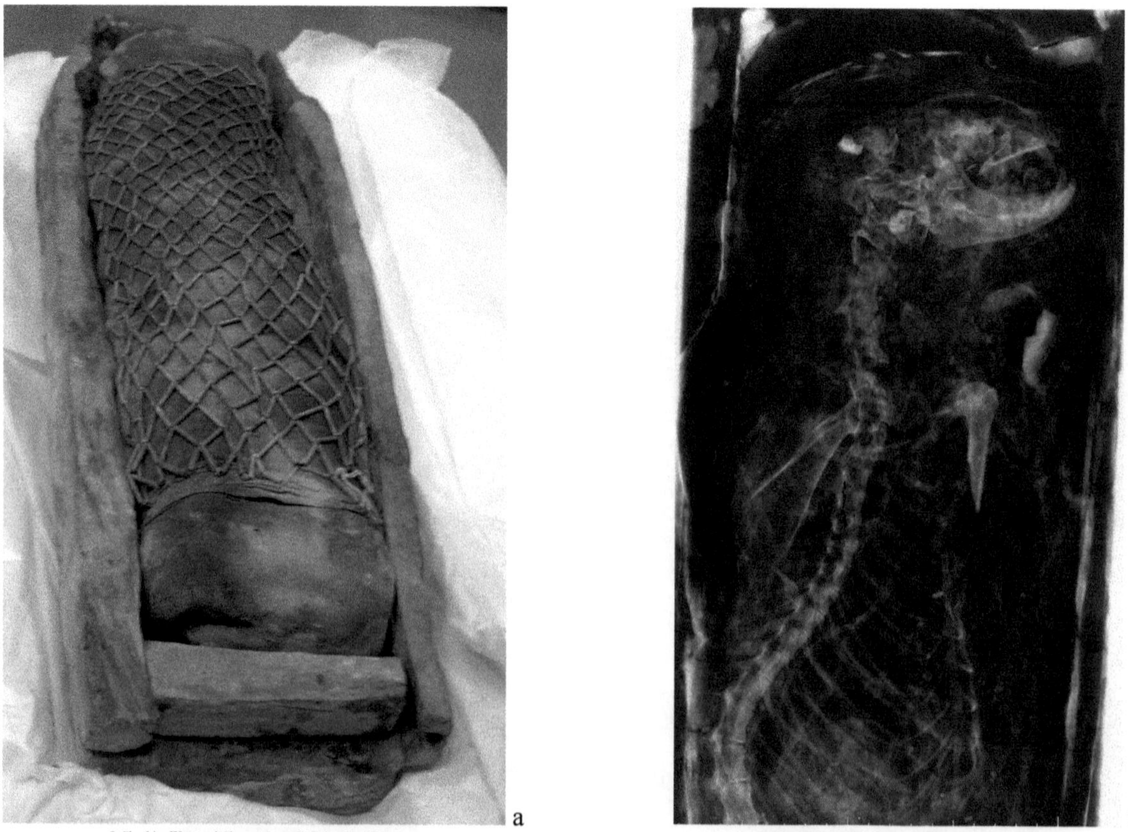

a © Cheshire West and Chester Council. Taken by J.E. Adams

b © Manchester Royal Infirmary / University of Manchester

Fig. 6.19 – a) Photograph and b) AP radiograph of mummy MA2.6 showing the pattern created by the beaded mesh which lies on top of the mummy itself

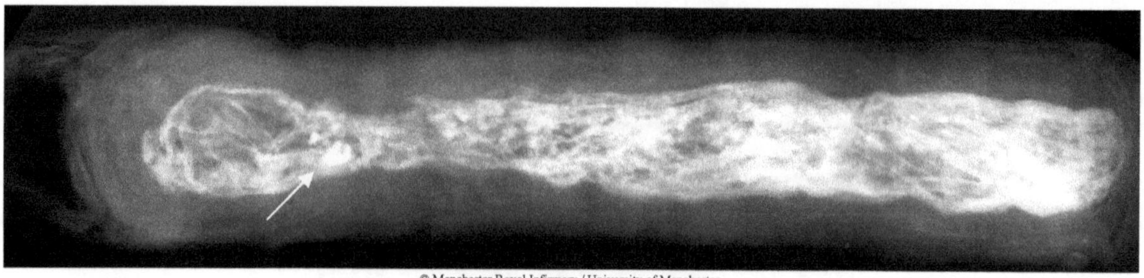

© Manchester Royal Infirmary / University of Manchester

Fig. 6.20 – Radiograph of mummy MA2.8 showing pellet-like opacities in the throat region

completely rule out the possibility that the embalmers may have placed material into the throat cavity during mummification. CT was unable to provide a clearer insight into the nature of the pellets which may have helped to establish the reasons behind their discovery.

Inclusions

Many of the mummified cats studied show evidence that resin and a tar-like substance were employed during the mummification process. Three of the seven disarticulated feline skulls studied clearly show that a layer of tar or resin has been applied directly to the skin of the animal. This layer has attracted sand particles which appear on the radiographs as small radio-dense specks. It is presumed that these skulls would have originally formed part of more complete animal mummies that have become detached over time. Further analysis using biomedical techniques would undoubtedly be beneficial in establishing the nature of the substance used to coat the body.

Mummy MA2.7 was found to contain a circular anomaly which is thought to represent a wax disc which is visible within the wrappings (Fig. 6.21).

6.3 – DOGS

External Appearance

Both of the mummified dogs studied during this research project have been crudely wrapped in brown linen bandages with no external decoration.

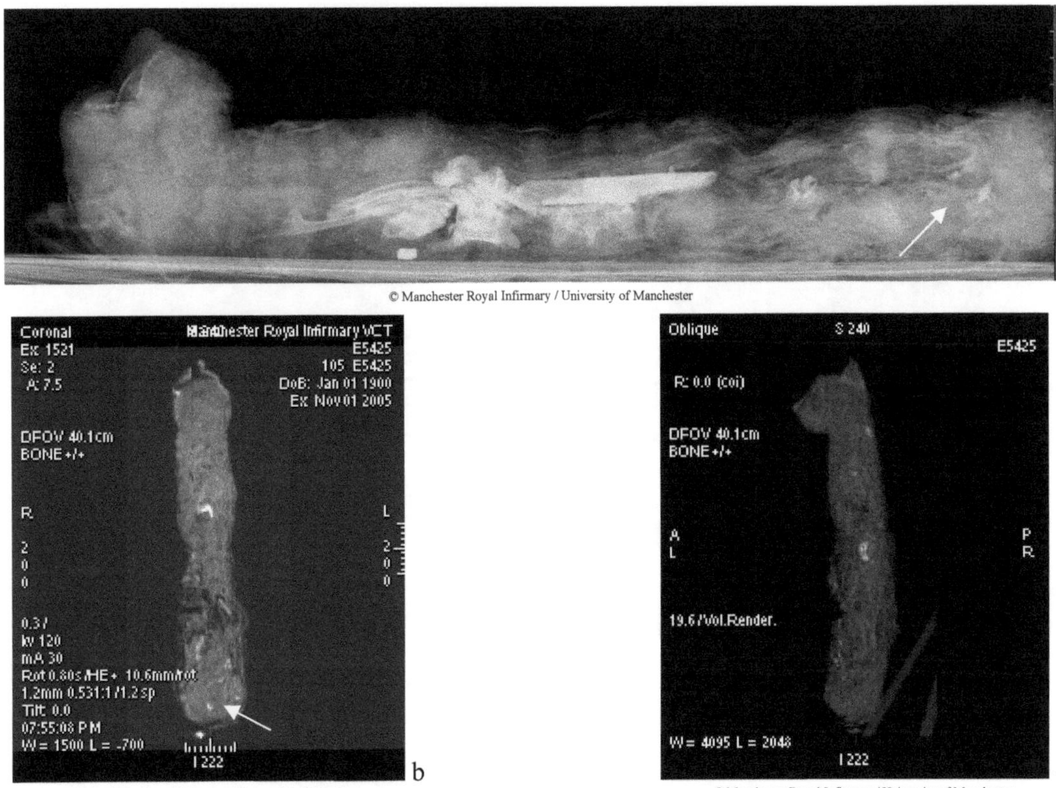

© Manchester Royal Infirmary / University of Manchester

© Manchester Royal Infirmary / University of Manchester

© Manchester Royal Infirmary / University of Manchester

Fig. 6.21 – a) Lat radiograph, b) Coronal reformation and c) Volume rendering image of mummy MA2.7 showing the presence of what is thought to be a wax disk

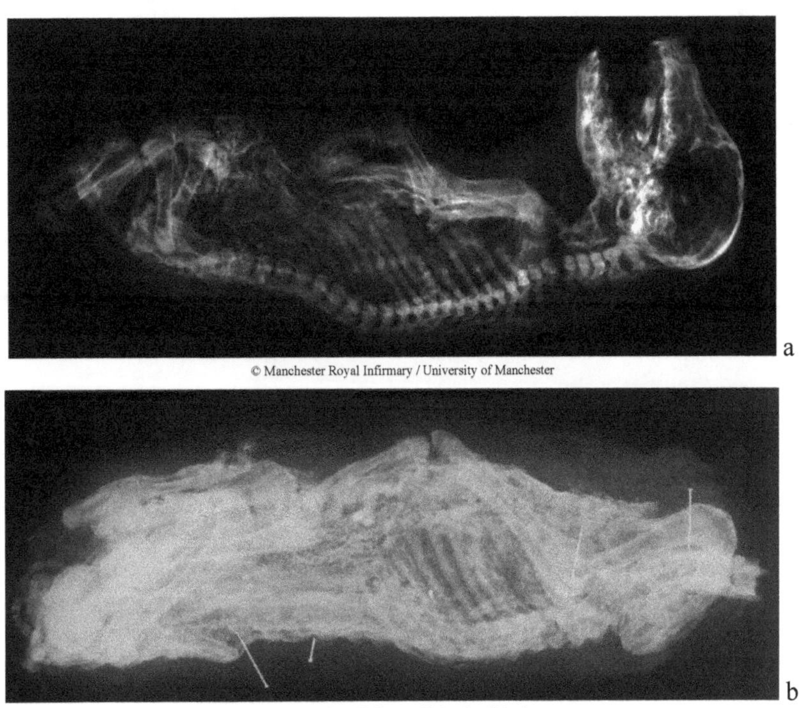

© Manchester Royal Infirmary / University of Manchester

© Manchester Royal Infirmary / University of Manchester

Fig. 6.22 – a) Radiographs of mummy MA3.1 and b) mummy MA3.2 showing the classic position of the skeleton

Mummy Shape

The dog mummies have been positioned into a cylindrical shape similar to that given to cats.

Position of the Skeleton

The two complete mummified dogs studied (MA3.1 and MA3.2) both appear to have been positioned in a similar

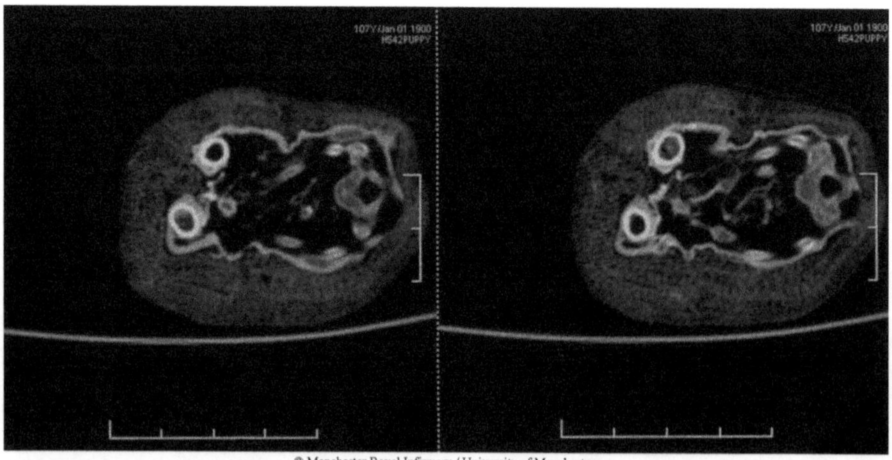

Fig. 6.23 – Axial CT slice images of mummy MA3.1 showing dense areas close to the skeleton of the animal, indicating that resin or natron could have been applied during the mummification process

way to cats with the upper limbs folded down and the lower limbs flexed upwards in front of the chest (Fig. 6.22). This seems logical as the animals are anatomically similar to each other.

Wrapping

Both of the mummified dogs studied appear to have been wrapped very tightly with the limbs folded close to the body. Radio-dense patches on the radiographs of MA3.1 suggest that natron or resin may have been applied to the animal's fur before wrapping began (Fig. 6.23).

Evisceration

The radiographic study did not yield any evidence to suggest that the dogs were eviscerated prior to embalming.

Packing of the Body Cavities

In the two dog mummies studied, none provided evidence to suggest packing of the body cavities.

Inclusions

A number of inclusions were identified through the radiographic process. Interestingly, mummy MA3.1 provides the only evidence to suggest the inclusion of false eyes within the orbital sockets witnessed in the entire study group (Fig. 6.24). This is not an elaborately decorated mummy, but is presented with plain brown bandages and no external decoration so the discovery on the radiographs was surprising. It should be noted that these anomalies were not noticeable on the plain film radiographs, only coming to light through the CT analysis.

Mummy MA3.2 demonstrates the speckled texture characteristic of sand or resin on the fur of the animal suggesting that the animal was treated prior to wrapping (Fig. 6.25).

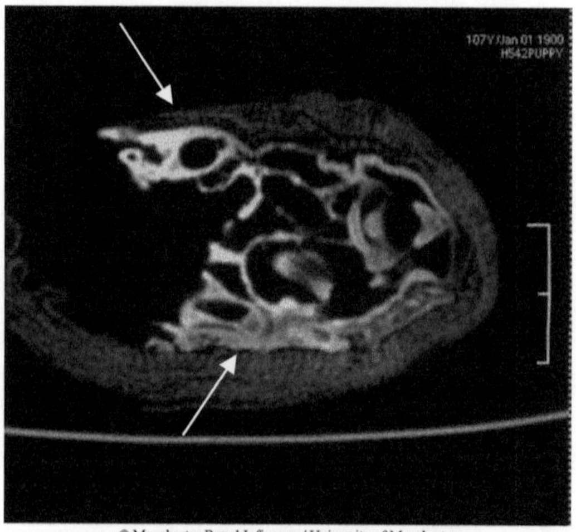

Fig. 6.24 – Axial CT slice of mummy MA3.1 showing inconclusive evidence that false eyes may have been applied during the mummification process

6.4 – FISH

External Appearance

Half of the fish mummies studied were not wrapped so they appeared simply as desiccated remains. A number of the wrapped mummies were crudely wrapped in brown linen. MA4.11 appears to have been constructed from reeds laid longitudinally along the length of the mummy before being given an incomplete layer of loose linen bandages. The notable exception is that of MA4.13 which is a small fish mummy wrapped in brown linen to which has been added exquisite external detail (Fig. 6.26). Facial features have been added using linen and paint in a stylised fashion.

Mummy Shape

The fourteen mummified fish studied have all been mummified in a natural position with no attempts being made to disguise their normal form.

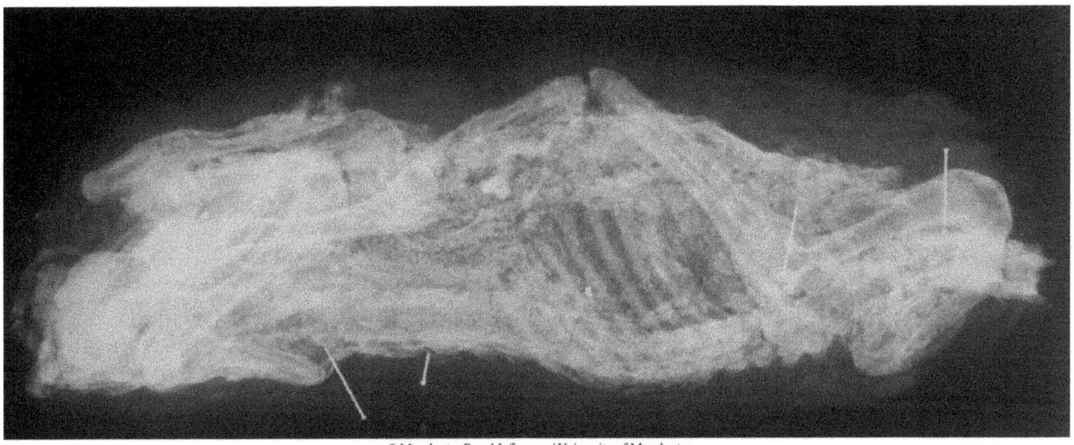

Fig. 6.25 – Radiograph of mummy MA3.2 showing a radio-dense substance close to the skin of the animal

a

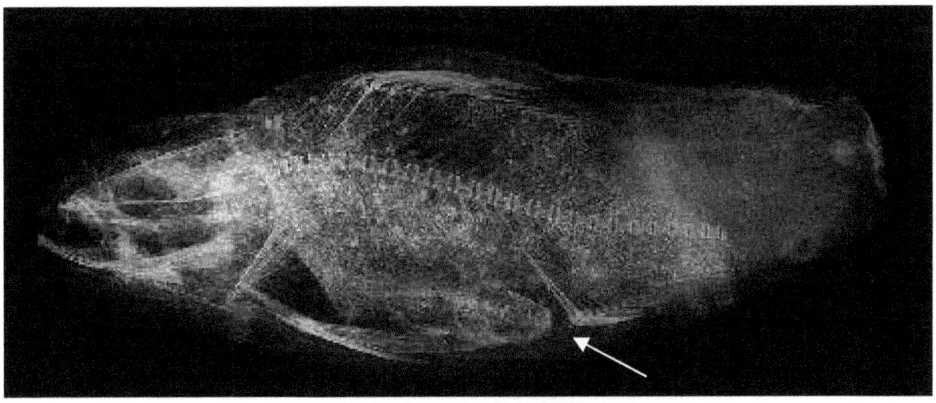

b

Fig. 6.26 – a) Photograph and b) AP radiograph of mummy MA4.13 showing the exquisite workmanship on the exterior of the bundle and the presence of a small fish in the gut area

Position of the Skeleton

All of the fish studied were found to have been mummified in a natural body position.

Wrapping

As mentioned above, seven of the fourteen fish mummies studied show no evidence of having been wrapped in any way. Mummy MA4.11 has been reinforced with reeds positioned along the full length of the fish which have then been bound with loose brown linen bandages.

The most elaborately presented fish mummy is MA4.13 which makes up in external detail what it lacks in physical size (Fig. 6.26). The mummy itself is tightly wrapped in plain bandages with great care being taken to add painted and modelled facial features. Upon radiographic analysis it was revealed that the embalmer had in fact added the face to the tail end of the animal.

Evisceration

Fish would have been easily eviscerated by making an incision along the belly of the animal and removing the

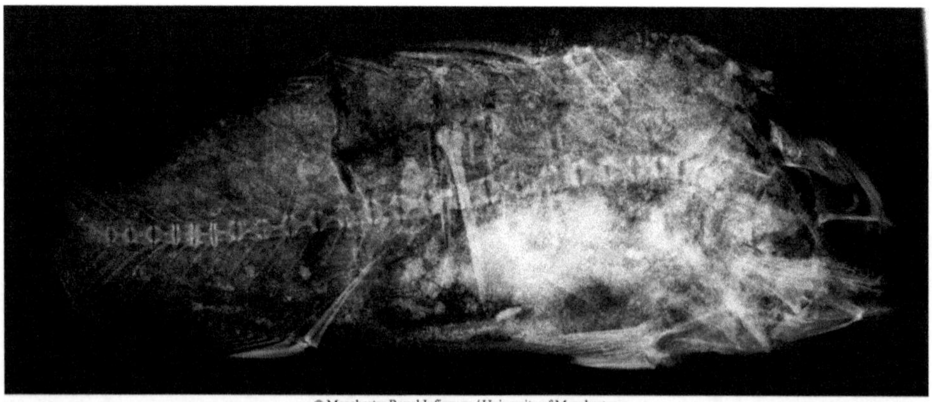

Fig. 6.27 – Radiograph of mummy MA4.2 showing the presence of a granular mass in the gut area

internal organs much in the same way that fish are gutted today. The resulting void could then be packed to fill out the body shape. No conclusive evidence was identified during the radiographic analysis to suggest that this practice had taken place. Mummy MA4.13 may have been eviscerated before a small fish was placed inside the body cavity (Fig. 6.26).

Packing of the Body Cavities

Two of the fish mummies studied (MA4.2 and MA4.12) showed evidence that the gut cavity contained a granular substance although it is thought to be likely that this represents ingested material (Fig. 6.27 and Fig. 6.28). Mummy MA4.5 shows evidence of small bones in the same area which adds to the theory that this represents ingested material. In the case of MA4.13, there is a small fish in the gut area which is thought to have been inserted into a slit made in the belly of the animal (Fig. 6.26).

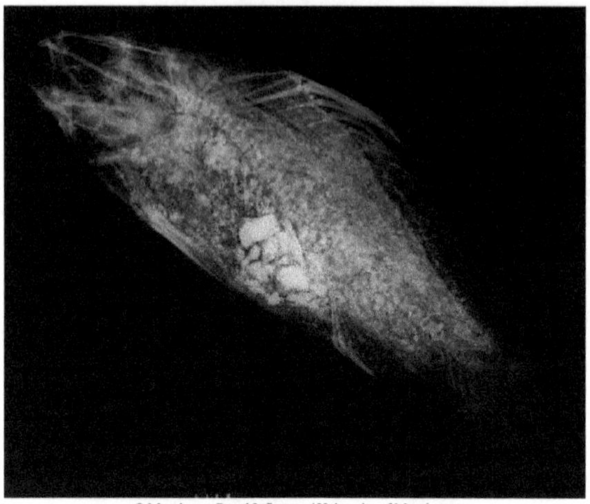

Fig. 6.28 – Radiograph of mummy MA4.12 showing the presence of granular material in the gut area

Inclusions

No evidence was found to suggest the presence of any intentional inclusions.

6.5 – CROCODILES

External Appearance

The crocodiles studied vary greatly in external appearance from unwrapped fragments to elaborately decorated mummies. Many showed signs of having been wrapped in reeds bound with thin thread. Bandages arranged in geometric designs had been used on two mummies – MA5.3 (Fig. 6.29) and MA5.7 (Fig. 6.30).

The only large crocodile mummy studied (MA5.4), estimated to be around five years of age at the time of death, was found to be completely unwrapped, but showed extensive evidence of the application of a tar-like substance to the exterior of the animal (Fig. 6.31). Interestingly, a number of the unwrapped juvenile crocodiles were found to have been treated in a similar way. MA5.5 has been given eyes made from linen on the external aspect of the mummy bundle (Fig. 6.32).

Mummy Shape

All of the crocodile mummies studied are presented in ways which reflect their natural anatomical body shape. One mummy (MA5.10) has been given individually modelled legs.

Position of the Skeleton

The majority of crocodile mummies studied display the same body position with the body outstretched and the limbs folded down close to the body. Three mummies (MA5.1, MA5.6, MA5.12) show signs of the limbs having been preserved in a more natural position away from the body (Fig. 6.33). This may have occurred if the animals were killed by dipping them in molten bitumen and the limbs being allowed to settle naturally.

Wrapping

The majority of the crocodile mummies studied during this research project are either unwrapped remains or they have been wrapped using reeds and thin thread. Mummy MA5.10 shows evidence that the natural body shape has

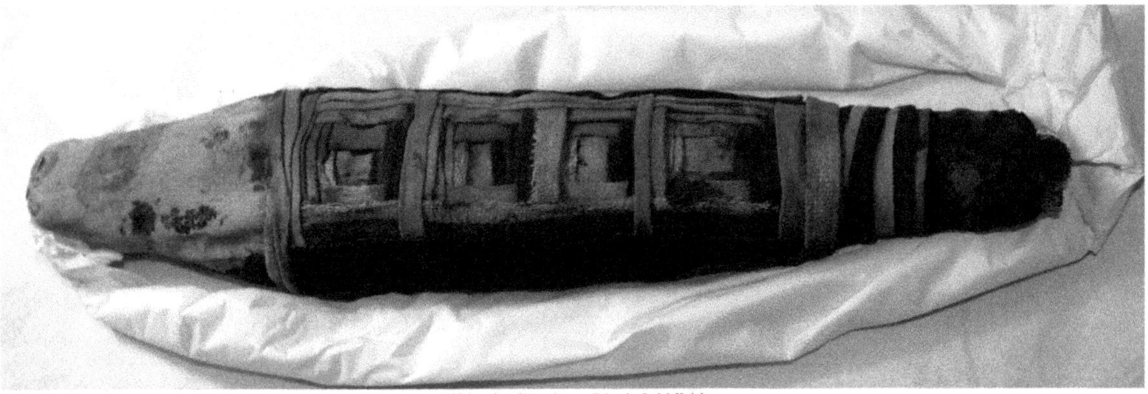

Fig. 6.29 – Photograph of mummy MA5.3 showing bandages applied to form a geometric pattern over the exterior of the bundle

Fig. 6.30 – Photograph of mummy MA5.7 showing geometric bandaging consisting of brown and white linen

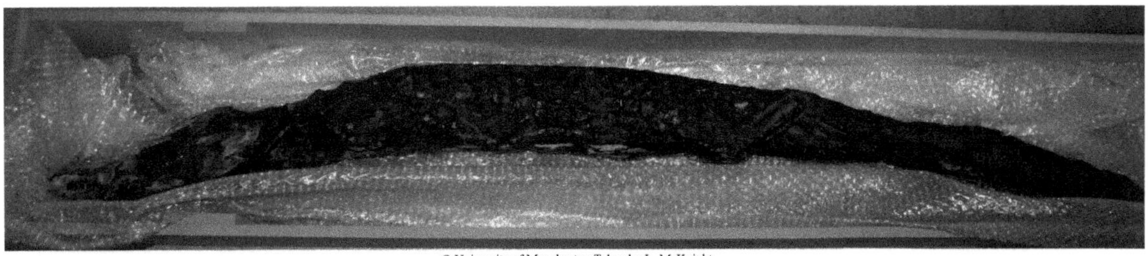

Fig. 6.31 – Photograph of mummy MA5.4, a completely unwrapped crocodile mummy showing evidence of a tar-like substance on the exterior

Fig. 6.32 – Photograph of mummy MA5.5 showing the addition of modelled eyes

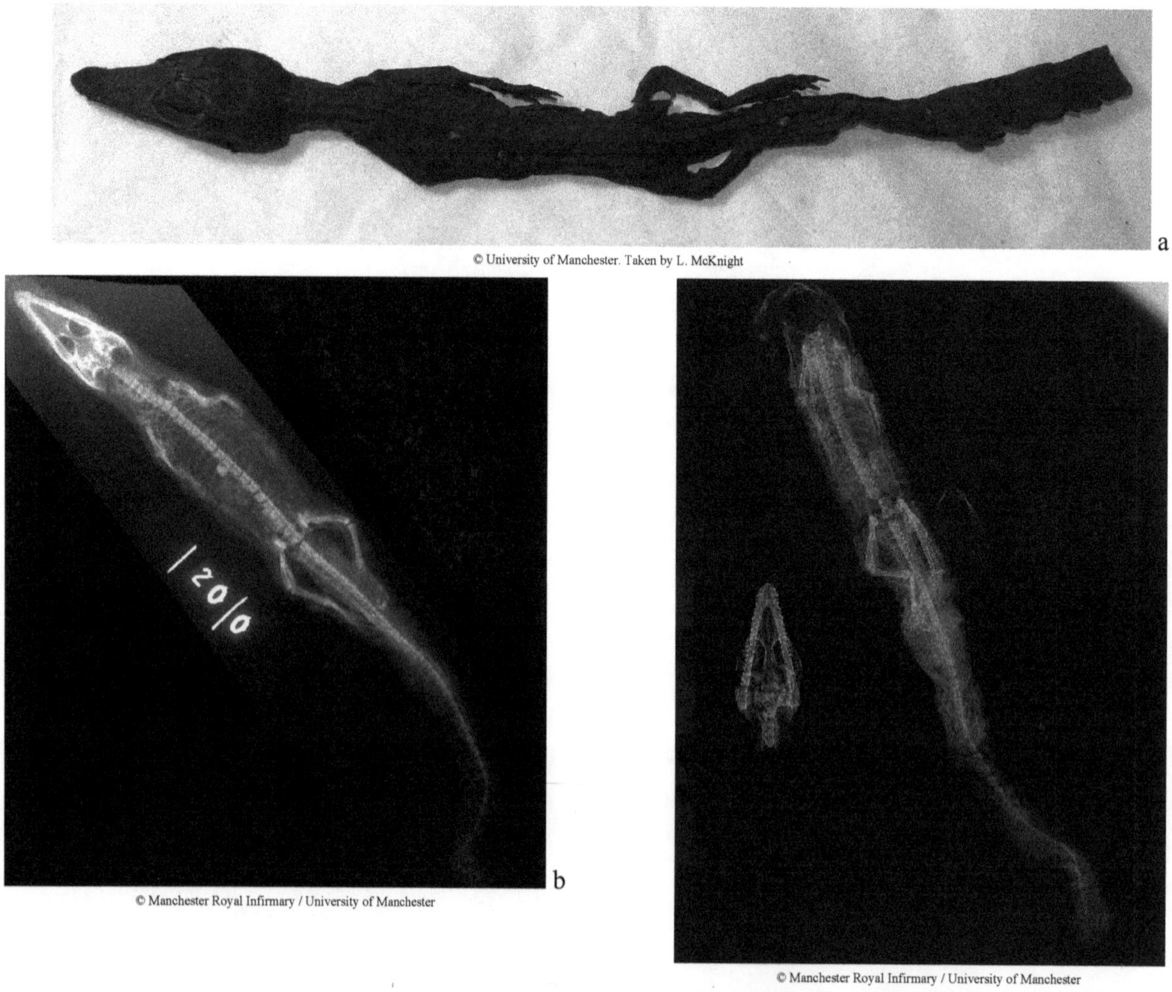

Fig. 6.33 – a) Photograph of mummy MA5.1, b) Radiograph of mummy MA5.6 and c) Radiograph of mummy MA5.12 all displaying the skeleton in a natural position

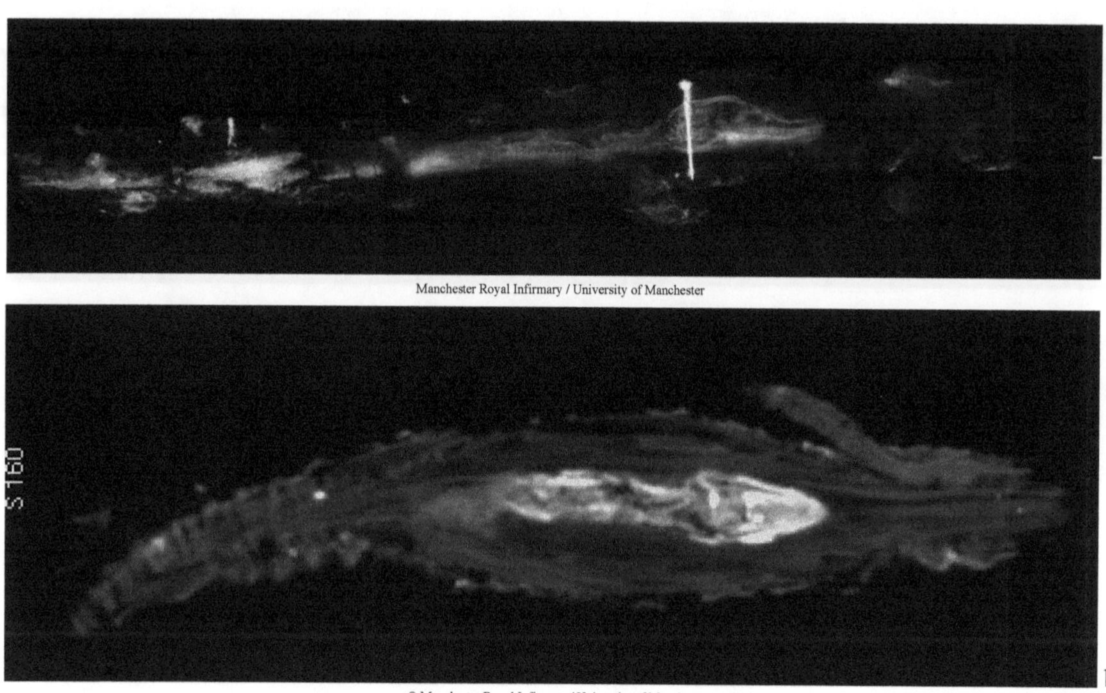

Fig. 6.34 – a) Lat radiograph and b) CT reformation of mummy MA5.10 showing the presence of reeds laid longitudinally to reinforce the body shape. Note how the CT image provides a much clearer insight into the nature of the contents

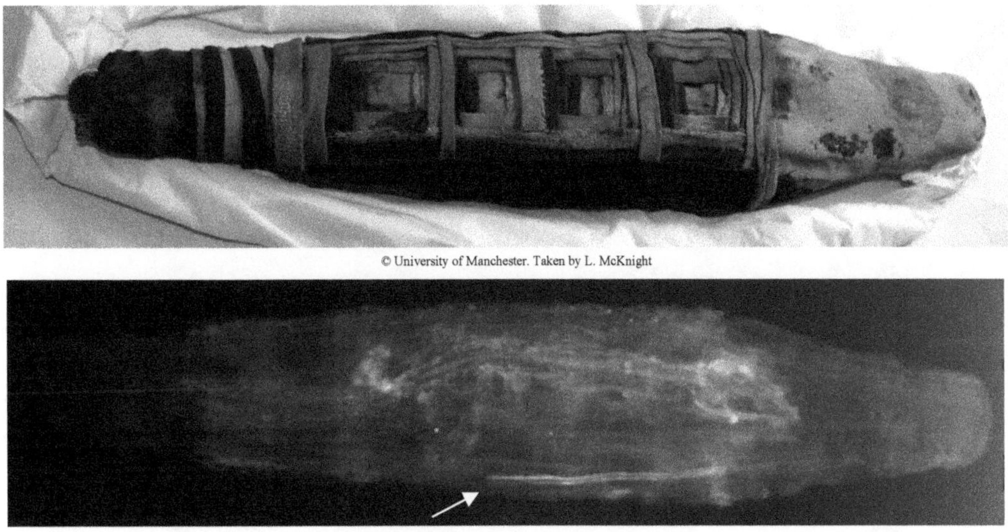

a

b

Fig. 6.35 – a) Photograph and b) AP radiograph of mummy MA5.3 showing the presence of a radio-dense inclusion, possibly a reed coated in metallic paint, tar or resin

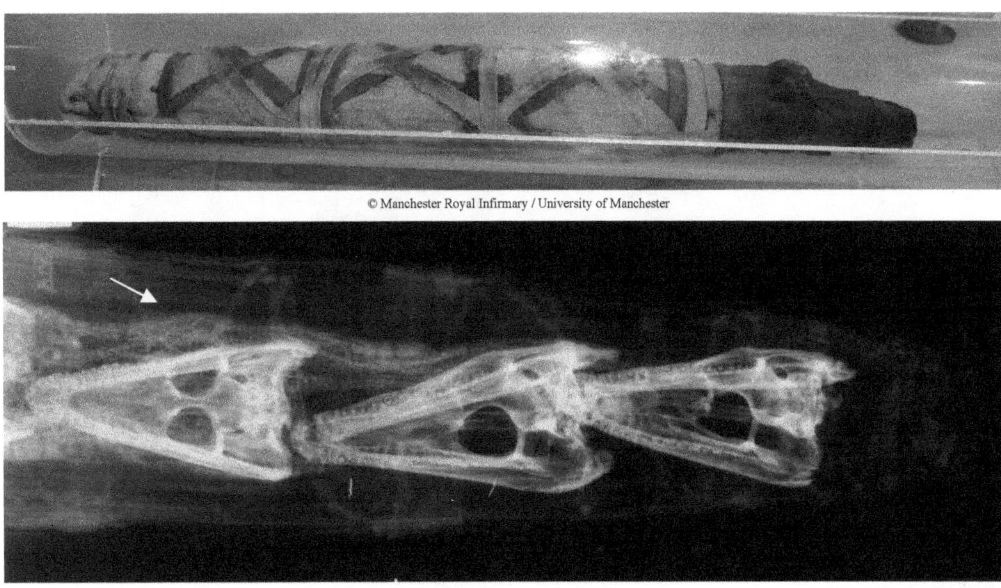

a

b

Fig. 6.36 – a) Photograph and b) Radiograph of mummy MA5.5 showing the presence of three crocodile skulls. A juvenile crocodile or snake appears to have been wrapped alongside the skulls

been accentuated using reeds and fabric to create the form of the limbs and tail (Fig. 6.34).

Evisceration

No evidence was found to suggest that evisceration had been practised on any of the crocodile mummies. There would almost certainly have been no point in eviscerating the juvenile animals due to their tiny proportions. The large crocodile (MA5.4) did not yield any evidence of the practice.

Packing of the Body Cavities

No evidence was found during this study to suggest that the body cavities were intentionally packed. A number of the crocodiles studied did show evidence of material within the body cavities; however, it was thought that this represented ingested material in all cases.

Inclusions

The radiographic analysis of mummy MA5.3 revealed an interesting anomaly (Fig. 6.35). The area in question appears on the radiographs as a region of radio-lucence with radio-dense margins. It is postulated that this may represent a reed coated in a substance such as metallic paint, tar or resin.

MA5.5 revealed surprising radiographic results (Fig. 6.36). The bundle was thought to be a complete crocodile mummy, but upon radiographic analysis was found to

a

b

Fig. 6.37 – a) Photograph and b) Radiograph of mummy MA6.3 showing the presence of a disarticulated juvenile snake skeleton within a crudely carved wooden box

contain three crocodile maxilla. To either side of the largest maxilla there appears to be the remains of a juvenile crocodile or snake.

Granular inclusions were visible on all the mummies which had been treated with the tar-like substance, presumably because it was attracted to the sticky substance before it dried.

6.6 – OTHER ANIMALS

This section refers to any species identified during the study which does not belong in any of the categories mentioned above. There were seven mummies in total which belong in this category.

External Appearance

A small number of mummies in the study collection were presented in boxes or sarcophagi of some sort. MA6.3, the mummified remains of a snake from the site of Hibeh dated to the XXVI dynasty, appears not to have been wrapped in bandages, but the animal had simply been laid in a crudely made wooden box with a sliding lid, on which was carved a snake motif (Fig. 6.37).

Mummy Shape

Mummy MA6.1 was recorded as being the mummified remains of a crocodile from the Roman cemetery site of Hawara. The outward appearance of the mummy bundle was the obvious reason for such an assumption being made as the embalmers had carefully modelled the form of a juvenile crocodile from reeds and light brown bandages. This carefully crafted form helped to disguise the fact that the bundle actually contained the remains of a juvenile snake found in a coiled position.

Position of the Skeleton

The three snake mummies studied show that the animals were placed naturally in a coiled or semi-coiled position.

Wrapping

One of the most intriguing specimens in this category is MA6.2 which has been tentatively identified as being the skeletal remains of an African giant shrew (*Crocidura flavescens*) or a house musk shrew (*Suncus murinus*) (Fig. 6.38). The mummy is plainly wrapped in brown linen bandages with the interesting addition of what appears to be a smaller package fixed to the top

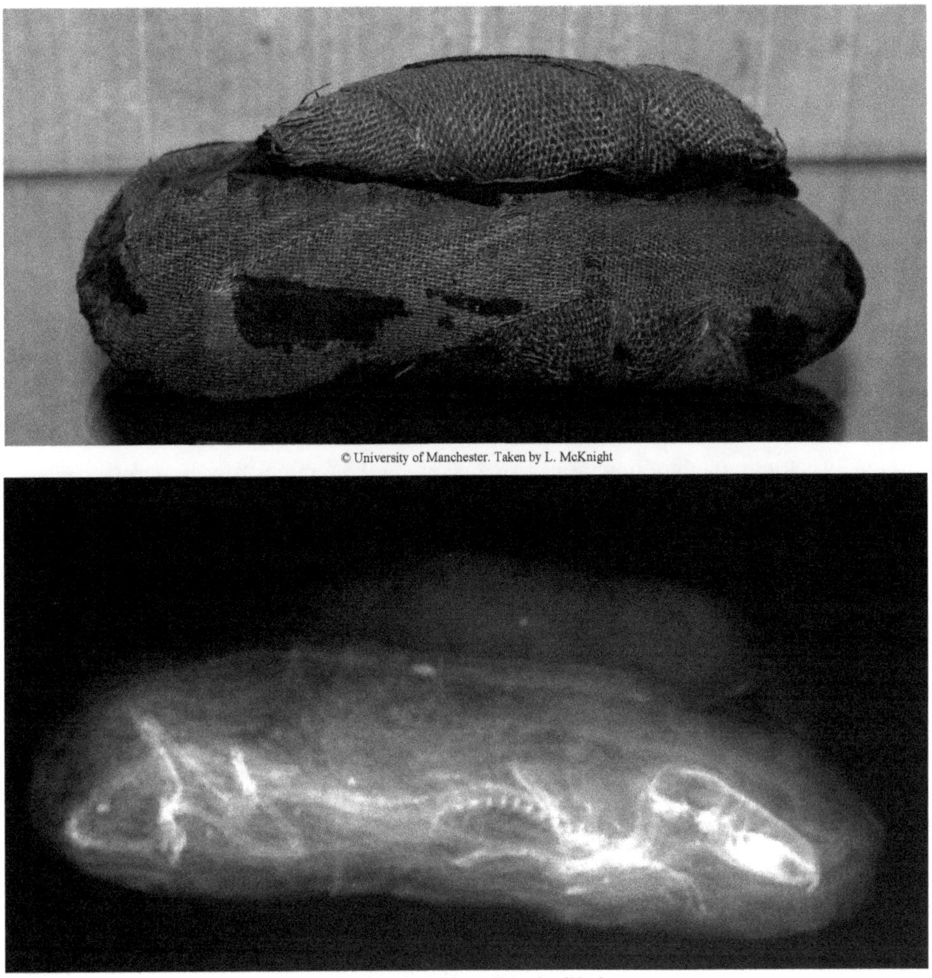

a

© University of Manchester. Taken by L. McKnight

b

© Manchester Royal Infirmary / University of Manchester

Fig. 6.38 – a) Photograph and b) Lat radiograph of mummy MA6.2 containing the skeletal remains of a shrew with the addition of a 'backpack' created from fabric which appears to contain a radio-lucent substance of unknown origin

resembling a 'back-pack'. Radiographs of this mummy clearly show the remains of the animal with the package containing a radio-lucent substance of an unknown nature. It is believed that this 'back-pack' may have been intended as a container in which food supplies were stored for use by the animal in the next world.

Evisceration

None of the specimens showed evidence of the practice of evisceration. All seven of the specimens discussed in this category were unlikely to have been eviscerated due to the nature of the species concerned (ie. snake) and their small dimensions.

Packing of the Body Cavities

No evidence could be identified to suggest packing of the body cavities.

Inclusions

The nature of the contents of the 'back-pack' fixed to the top of MA6.2 would prove to be an interesting area for further study and would undoubtedly help to clarify the reasons behind the unusual appearance of the mummy bundle (Fig. 6.38). It has been suggested that the radio-lucent contents visible on the radiographic images could be grain particles.

The radiographic study of mummy MA6.3 (the snake presented in a wooden box) shows evidence of a radio-opaque mass surrounding the skeletal remains (Fig. 6.37). The skeleton appears to have become disarticulated as it decomposed so it is likely that no mummification treatment was performed. The animal may have simply been deposited within the wooden box along with a substance consisting of sand or mud. The fragmentary nature of the remains is similar to MA6.5 where the vertebrae have also become disarticulated.

Mummy MA6.7 was recorded as a mummified egg of unknown origin (Fig. 6.39). The bundle was not wrapped and it was clear that the egg had become compressed over time leading to the complete destruction of its structural integrity. The radiographic study of the mummy revealed that the egg contains the foetal remains of a chicken.

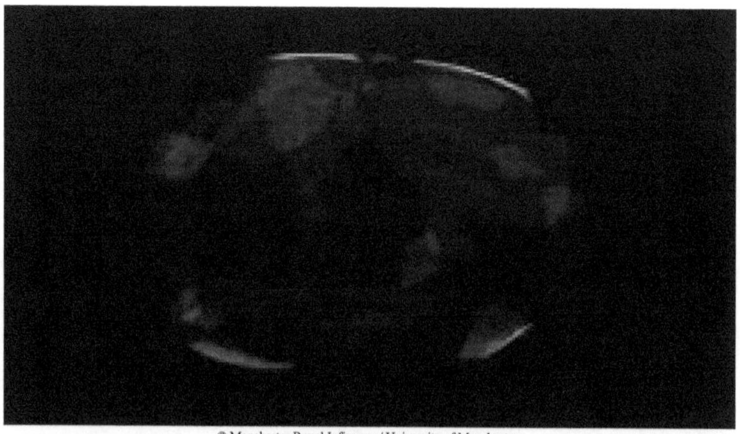

Fig. 6.39 – Radiograph of mummy MA6.7 showing the remains of a compressed egg

6.7 – DISCUSSION

Overall, the use of radiography and CT has been useful in helping to provide supplementary information on the nature of the mummification techniques employed by the ancient Egyptians in their treatment of animal remains. The issues arising during the study, and those which prevented conclusive answers to the research questions being identified, are believed to be due to the nature of the mummified animal remains available for study rather than to the choice of the techniques used. This was hypothesised prior to the radiographic examination and was found to be true in the case of a large number of the specimens.

These adverse issues revolve around the following facts:-

1. The mummified animals studied all belong to species groups which are either naturally small in proportion or they are juvenile animals. This means that the information gathered through the radiographic process is limited by spatial and contrast resolution factors which limit visual examination of the skeletal remains themselves. Size of the animal has been noted in the sections relating to evisceration and the packing of the body cavities as a potential reason for why the embalmers may have considered the practice to be counter-productive.

2. The degree of compression caused by the application of wrappings on the mummy can lead to skeletal deformity and consequently the misleading interpretation of the radiographic images.

3. A poor state of preservation of both the skeletal remains and the mummy bundle can cause problems with interpretation and the ability to identify potentially interesting information on the mummification techniques employed. In many cases, the animal remains show signs of extensive decomposition which may have occurred prior to embalming or as a result of poor, or non-existent, embalming techniques. This would lead to degradation causing fragmentary and disarticulated remains with little or no skeletal integrity to survive within the bundle.

In my opinion, had the study group contained a greater proportion of larger animal mummies, it would have been possible to report more conclusive findings relating to the mummification techniques. This is because larger animals would allow greater scope with which to assess the potential of radiographic analysis techniques in ascertaining how the body was treated following death. As mentioned above, interesting results have been gathered from human and large animal mummies studied previously in this manner. It is therefore possible to report that the radiographic techniques employed during this study were unable to provide any direct evidence for techniques such as evisceration and the removal of the brain.

This study highlighted the potential importance of a thorough external examination of the mummies. External observations help to provide the researcher and radiographer with a first-hand impression of the mummy considering the following factors:-

1. The state of preservation – it is vitally important to assess the degree of preservation of each mummy prior to the radiographic analysis to ensure that the specimen is treated in the most suitable manner. Many of the specimens which are presented in boxes or containers have been placed there for a reason and often because they are simply too fragile to be lifted without this additional support. All specimens which showed signs of being poorly preserved or excessively fragile were imaged in their containers and this was not found to be detrimental in any way to the resulting images.

2. The one notable exception to this is the ibis mummy MA1.20 which was transported to the study location in a plastic box filled with vermiculite beads to protect its fragile composition. The vermiculite gives the resulting images a snowy appearance and makes interpretation difficult. However, the point must be made that this conservation technique was important

as without it the object would not have been allowed to leave the confines of the museum and no images would have been gathered.

3. The presence or absence of wrappings can be initially determined through analysis of the external appearance of the mummy bundle. This information can then be used in collaboration with the radiographic images to enable a more detailed assessment of the wrapping techniques employed. CT was found to be particularly beneficial for analysing the wrapping techniques. For example, many wrapped mummies were shown on the radiographic images to have been wrapped in stages which would not have been ascertained through an external examination alone.

4. External examination helped to identify that the embalmers appear to have used standardised body shapes when manufacturing votive animal mummies. This may have been brought about through force of habit when creating thousands of similar offerings or there may have been ritual significance to certain body shapes. It seems realistic to suggest that practical reasons played a large part in this with the natural body shape of the animal in question pre-determining the eventual form of the mummy bundle.

The position of the skeleton appears to follow the standardised and, in many cases previously documented forms, in the majority of cases. It is the exceptions to these forms which provide interesting contrasts and enable discussions about the practices employed by the embalmers. In some cases it appears that they made life difficult for themselves by increasing the amount of work that it would require to manufacture a mummified animal. It is believed that some of the more unusual skeletal positions identified may have been the result of heavy handed treatment of the remains, a lack of diligence, or the simple experimentation on the behalf of the embalmers. It is possible that animal mummification provided a means through which the embalmers could test new techniques which would have been forbidden on human remains.

It is possible that some animals were not mummified immediately following death. If rigor mortis had been allowed to set in, it is possible that the body was already rigidly fixed when the animals reached the embalmer's workshop. This may have been the case with animals that died of natural causes and were not discovered immediately. In these cases it seems logical that the animals would have been preserved in the position in which they reached the place of mummification.

One mummy which demonstrates this theory is the only confirmed modern mummy to have been studied in this project. MA6.4 is the naturally mummified remains of a rat found in a local chimney (Fig. 6.40). It was known at the time of study that this mummy was neither ancient nor Egyptian, but it was decided to take the opportunity to analyse it using radiography to use as a comparison with those specimens believed to be ancient Egyptian in origin. The rat has been excellently preserved almost as if asleep in a curled position with the tail folded along the spine.

The presence or absence of inclusions within the mummy bundle can only be identified through the use of radiographic techniques without facilitating the destruction of the specimen completely through dissection. It is with this particular research question in mind that the potential of radiography really becomes evident.

Numerous mummies studied during this research project were found to contain inclusions of some form (either intentional or unintentional) ranging from sand particles and granular material through to oddities such as isolated human bones and suspected amulets. The presence of these items would not have been demonstrated without the use of these non-destructive techniques.

The combination of the two radiographic techniques used during the study – X-ray and CT – appear to compliment each other in terms of the information that they are able to provide to the researcher. Radiography is the first method of study which should be employed to give the preliminary images of the mummy and the contents within. In many cases CT will elaborate on the information gained from X-ray and enable a more complete and thorough analysis of the mummy to be carried out. The use of CT proved to be exceptionally valuable in the identification of inclusions and the wrapping techniques employed in the manufacture of the mummies. It was less helpful in this study in answering questions relating to evisceration and the packing of the body cavities; however this is thought to be due to the nature of the mummified material available rather than a limitation of the technique.

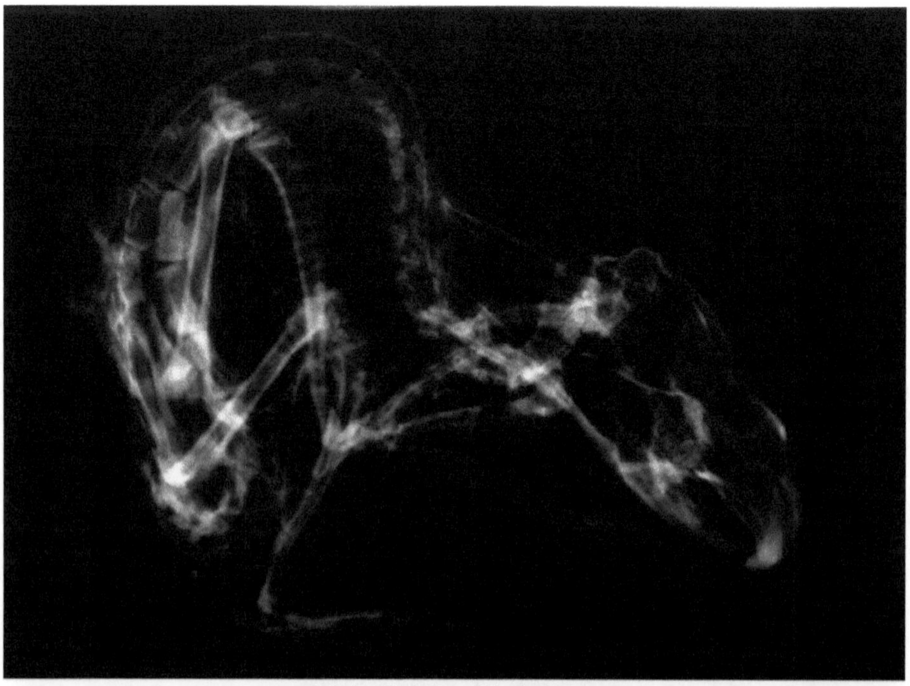

a

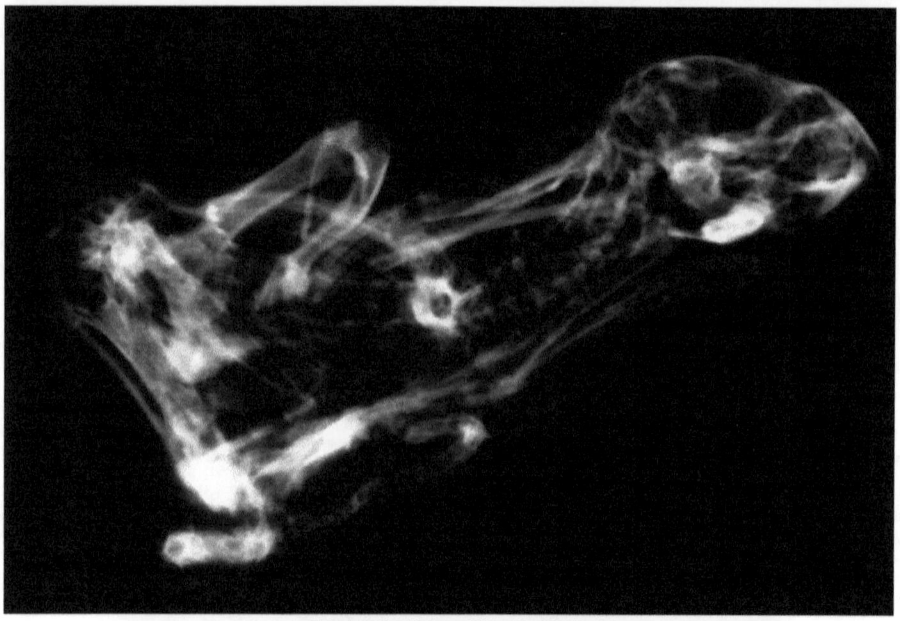

b

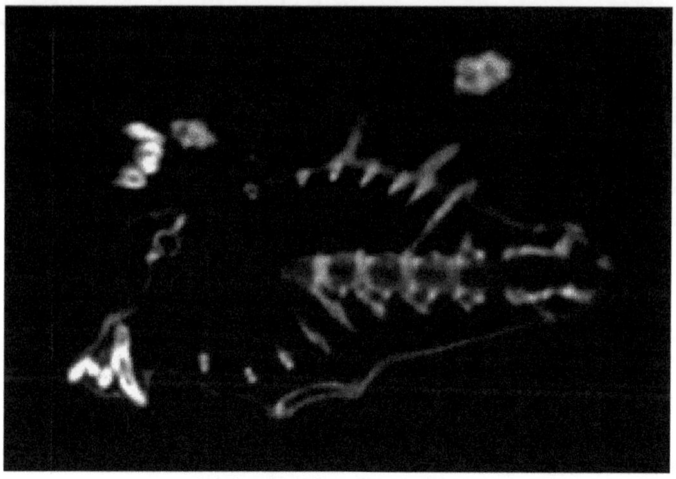

c

Fig. 6.40 – a) AP radiograph, b) Lateral radiograph and c) CT axial section of mummy MA6.4, a naturally mummified modern rat, showing excellent skeletal preservation

Chapter Seven
PSEUDO-MUMMIES

7.0 – INTRODUCTION

> "Some of the animal mummy bundles examined turned out to be ancient fakes, which reveal a very human aspect of the ancient Egyptians"
>
> Ikram and Iskander commenting on pseudo-mummies identified during their research at The Egyptian Museum, Cairo (2002, IV)

In total twenty-eight of the 127 mummies (equating to twenty-two percent of the study group) studied during this research project were found upon radiographic examination to be 'pseudo-mummies'. This term is used here to describe any mummy which was found to contain no identifiable animal skeletal material. These pseudo-mummies present an area of interest to the Egyptologist as they present something different to the 'normal' standard of animal mummies. They are unique and the reasons for their manufacture are largely unknown.

7.1 – BODY FORM

The absence of skeletal material within the pseudo-mummies must have presented a unique challenge to the embalmers whose task it was to produce objects which would appeal to the visiting pilgrims. The skeleton would have acted as a framework around which the mummy bundle could be created; without this, it would have been necessary to form the body shape from another substance to the point that it resembled an animal. The radiographic analysis suggests that there were a number of ways in which it would have been possible to produce this.

In one particular case (MA7.4) the body shape had been constructed from bandages alone simply rolled into a coil and formed into a circle (Fig. 7.1). This was presumed from its external appearance to be that of an eel, but was found upon examination to contain only bandages with no skeletal material whatsoever. What this mummy was in fact intended to represent remains unclear.

In the majority of cases it appears that the body shape would have been created from a mud-like substance. This seems logical in that the embalmers would have been able to access sources of mud which would have been easily moulded into the desired form and would have dried out over time to create a hardened mass. Radiography often shows radio-dense areas within this mud-like substance which may represent naturally occurring sand or stone fragments.

Pseudo-mummies intended to resemble crocodiles appear to have been consistently produced using reeds laid longitudinally to create the elongated shape of the body. This also appears to have been the case with actual crocodile mummies where the reeds have been used to reinforce this shape and provide additional support to the skeleton. Generally in these cases, the use of reeds negates the need to use bandages and so the reeds are simply bound with thin thread to hold their shape.

The radiographic analysis of the pseudo-mummies revealed evidence of air filled voids within a number of the mummies. This is probably due to the contents settling over time especially in the case of mummies such as MA7.18 which have little, or no, structural integrity with which to hold them together (Fig. 7.2). In some cases these voids may be the result of fractures occurring through the brittle bundles into which air has become trapped. This seems to have been the case with MA7.11 where air has entered the mid-section of the mummy bundle appearing on the images as dark areas. Reeds and feathers can often appear on the images as linear air-filled structures due to their hollow nature. Areas where the bandages have been loosely applied can often show air voids as in the case of MA7.6 where a layer of air has become trapped between the object and the wrappings.

Semi-calcified eggs were visible within a number of mummy bundles, either wrapped alone in the case of MA7.20 or alongside additional material (Fig. 7.3). It has not been possible to attribute them to a species, but it seems likely that they were viewed as sacred due to their

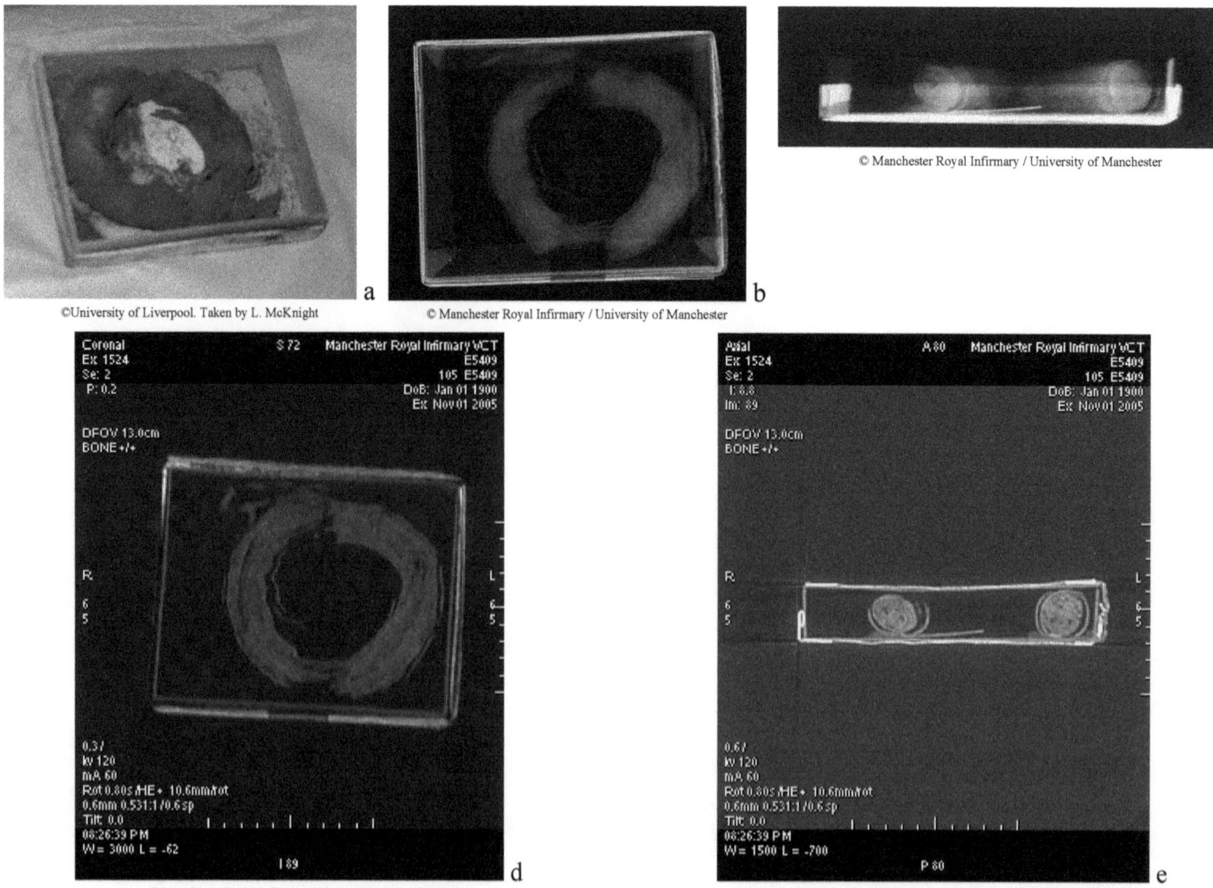

a ©University of Liverpool. Taken by L. McKnight

b © Manchester Royal Infirmary / University of Manchester

c © Manchester Royal Infirmary / University of Manchester

d © Manchester Royal Infirmary / University of Manchester

e © Manchester Royal Infirmary / University of Manchester

Fig. 7.1 – a) Photograph, b) AP radiograph, c) Lateral radiograph, d) Coronal reformation and e) Axial reformation of mummy MA7.4 created from coiled bandages

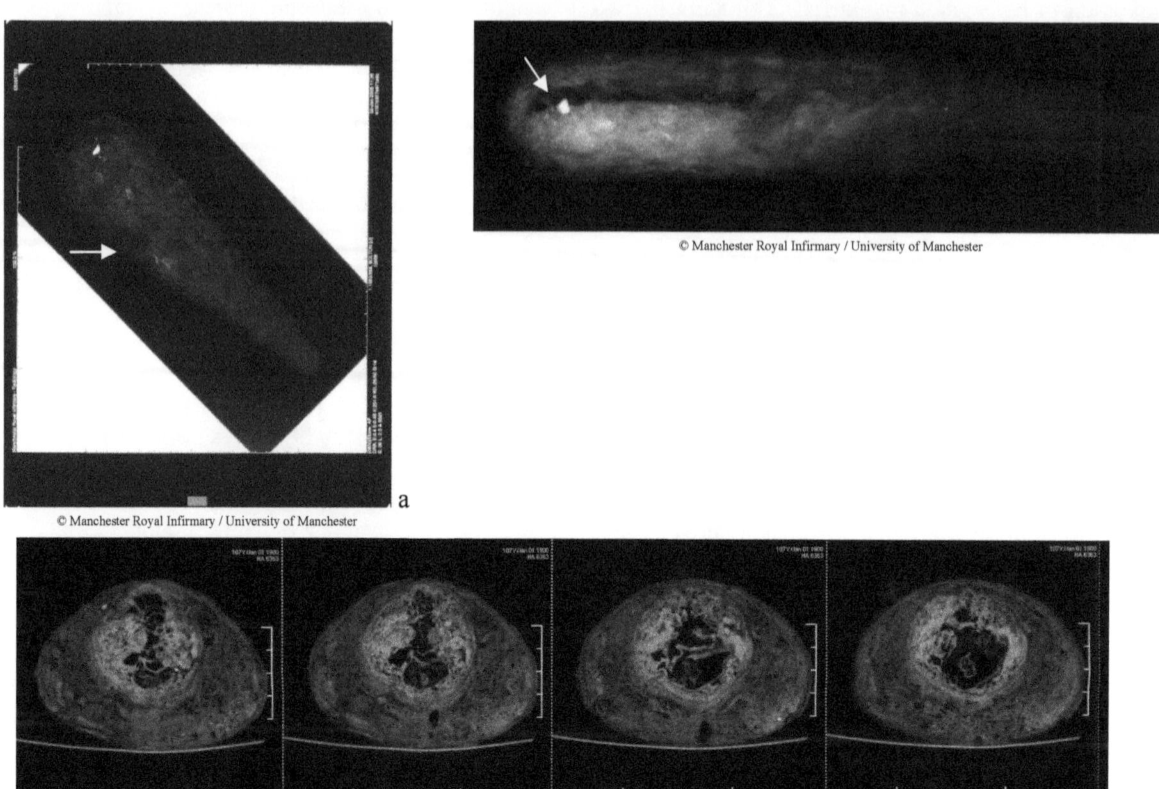

a © Manchester Royal Infirmary / University of Manchester

b © Manchester Royal Infirmary / University of Manchester

c © Manchester Royal Infirmary / University of Manchester

Fig. 7.2 – a) AP radiograph, b) Lateral radiograph and c) CT axial sections of mummy MA7.18. There is evidence of an air-filled void in the centre of the bundle where the contents have settled over time. Image c) shows the void clearly

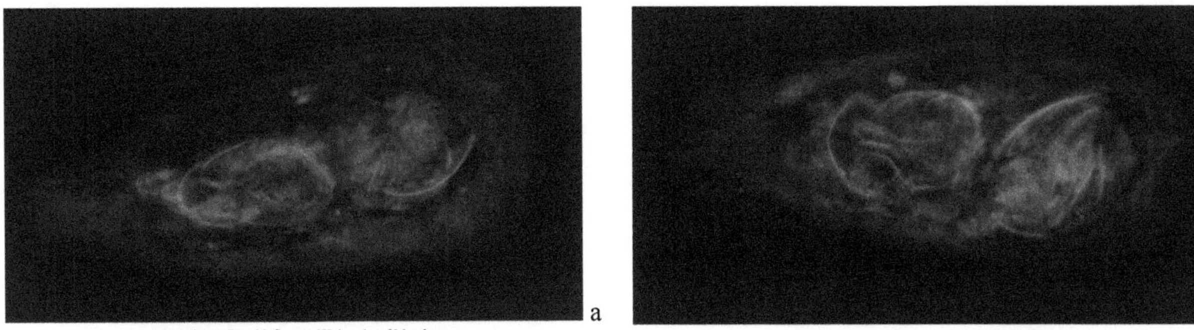
a b

Fig. 7.3 – a) AP radiograph and b) Lateral radiograph of mummy bundle MA7.20 which was found to contain what are thought to be two semi-calcified eggs

a b

Fig. 7.4 – a) radiograph and b) region of interest on mummy MA7.26 which was found to contain what is believed to be an open-rooted animal tooth. A metal pin is visible close to the region of interest which is a modern conservation device

association with a revered animal. They become naturally compressed during the mummification process but can be viewed as calcified anomalies on the radiographic images.

A number of intriguing contents were highlighted by the radiographic process. Mummy MA7.26 appears to contain a large animal tooth with an open root thought to belong to a pig, jackal or hyena within the mummy wrappings (Fig. 7.4). This bears no obvious connection to its crocodile form. Mummy MA7.27, a pseudo-mummy again intended to represent a crocodile, was found to contain a human phalanx bone (Fig. 7.5). How and why this bone came to be incorporated into what had been manufactured to closely resemble a crocodile remains unclear.

CT was successful in highlighting regions of interest in the case of mummies which demonstrated unusual anomalies. One particular example is MA7.25, a pseudo-mummy formed in the shape of a crocodile (Fig. 7.6). Two dense objects were visible on the images which were then analysed further using advanced computer programmes intended for medical diagnosis purposes. These programmes enable regions of interest to be analysed in more detail to reveal the attenuation of the anomalies. In the case of animal mummies, this technique can be used to attempt to define what material a structure is made from. The two dense objects visible within MA7.25 showed attenuation values of as high as 1253.6 which is similar to that of cortical bone or metal. This is a particularly good example of how the technological advances of CT can be harnessed to provide information on archaeological material.

7.2 – WRAPPING TECHNIQUES AND EXTERNAL DECORATION

The pseudo-mummies examined during the course of this project appear to have been wrapped in similar ways to those which contain skeletal material reinforcing the theory that they would have been produced by the same embalmers alongside legitimate animal mummies. Many show signs of having been treated with a tar-like substance which has attracted granular particles which have adhered to it during the mummification process.

It appears that a number of the mummies were wrapped in different stages. This appears on the radiographs in the form of distinct layers of bandaging, often with a tight initial layer before having a looser layer applied, and then finished off with a tight layer once again as in the case of MA7.20. The reasons behind this practice are unclear; however this was witnessed in a number of mummies which were found to contain skeletal material (see Chapter Six).

The radiographic analysis often highlights areas of radio-density which could represent the use of metallic gilding or paint. In many cases this appears as patches within the form of the mummy bundle itself, but it also appears as an external feature on mummy MA7.9 in the form of an elaborate gilded head and breastplate (Fig. 7.7). This specimen has been carefully modelled from a homogenous substance to resemble a hawk before being wrapped in brown linen bandages and adorned with gilding.

A variety of bandaging styles can be witnessed from this collection of pseudo-mummies. These range from plain

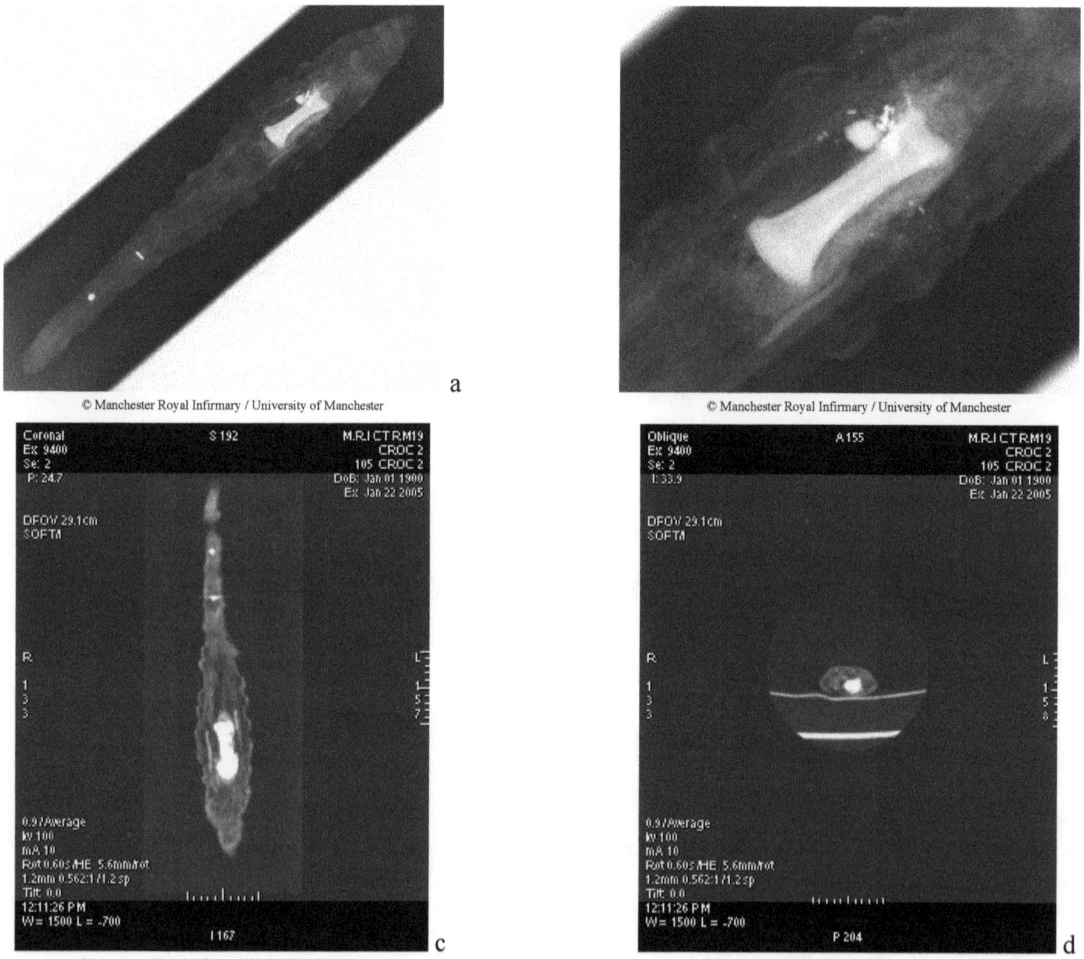

a © Manchester Royal Infirmary / University of Manchester

b © Manchester Royal Infirmary / University of Manchester

c © Manchester Royal Infirmary / University of Manchester

d © Manchester Royal Infirmary / University of Manchester

Fig. 7.5 – a) AP radiograph, b) Region of interest, c) Coronal reformation and d) Oblique reformation of mummy MA7.27, a pseudo-crocodile mummy containing a human phalanx bone. Image c) clearly shows the presence of reeds forming the shape of the crocodile

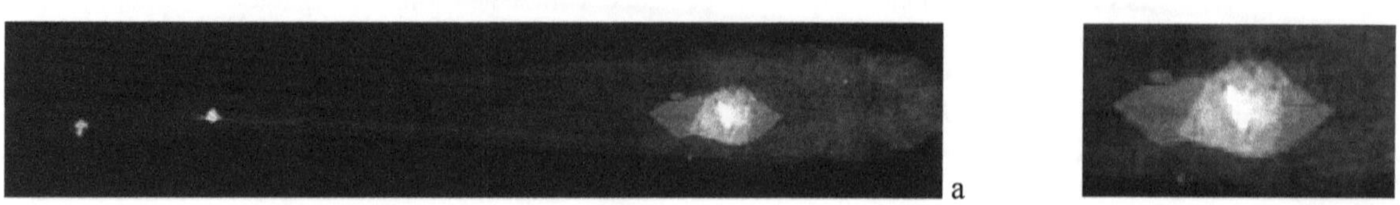

a © Manchester Royal Infirmary / University of Manchester

b © Manchester Royal Infirmary / University of Manchester

Fig. 7.6 – a) AP radiograph and b) region of interest on mummy MA7.25 showing an area of high attenuation

linen bandaging with no external decoration to elaborately wrapped mummies displaying bandages of different colours arranged to form patterns of concentric squares or diagonal lines. There can be little doubt that the embalmers lavished high levels of attention to detail on these mummies whether they contained animal material or not.

The addition of false ears, eyes and other facial features modelled from linen and applied to the external aspect of the mummy bundle was a relatively common occurrence within the study group. Paint has also been used on a number of specimens to show facial features which are often depicted in a somewhat stylised manner. This must have been intended to make the 'animals' appear more lifelike and to appeal more to the prospective purchaser. These methods appear to have been used on mummies containing skeletal material as well as pseudo-mummies.

7.3 – DISCUSSION

Relatively little evidence survives to support the work of the embalmers at the temple complexes devoted to the animal gods and to explain who the embalmers actually were. Travellers to Egypt often remark on the strange spectacles they witnessed; however, these are simply subjective comments and cannot be seen as an altogether accurate portrait of the industry. The travellers generally had little or no understanding of the belief system which underpinned animal mummification and as such showed little compassion for the sensitivity of the views of the

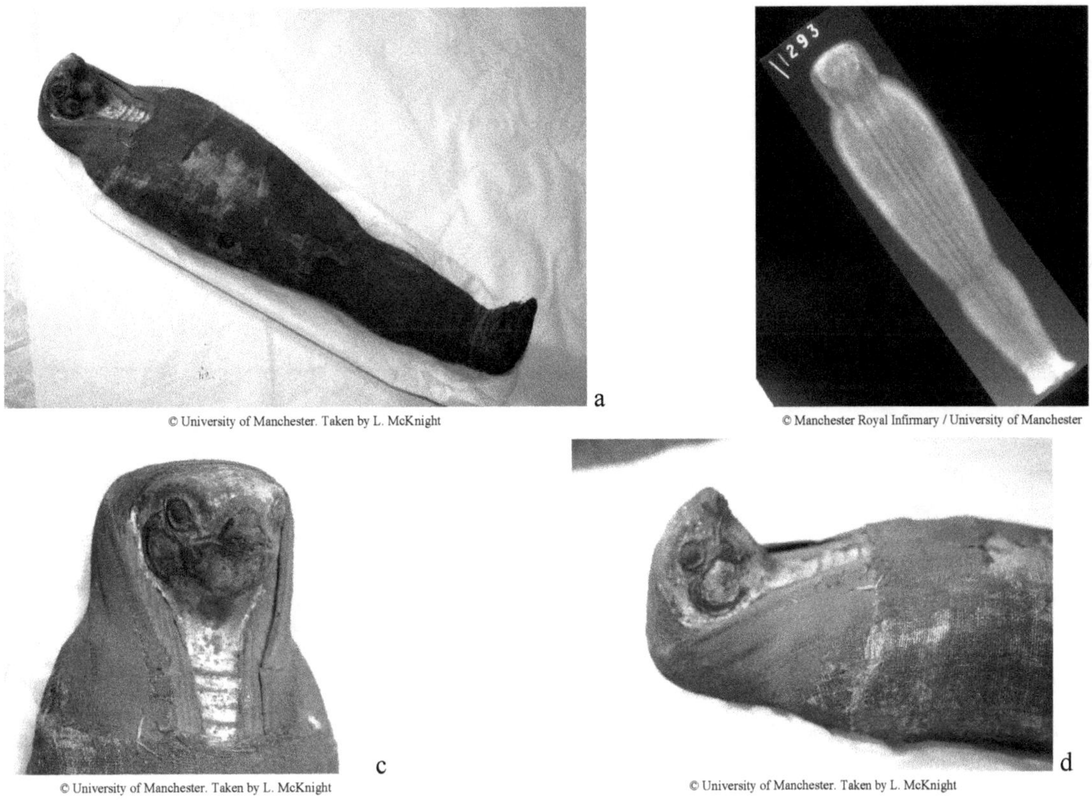

© University of Manchester. Taken by L. McKnight

© Manchester Royal Infirmary / University of Manchester

© University of Manchester. Taken by L. McKnight

© University of Manchester. Taken by L. McKnight

Fig. 7.7 – a), c) and d) photographs of mummy MA7.9 which demonstrates exquisite gilding to the face and chest area, b) AP radiograph demonstrating that the mummy bundle contains no human or animal skeletal material

ancient Egyptians. This means that Egyptologists and archaeologists rely heavily on the material remains relating to this practice and the burial contexts in which the mummies are found.

Pseudo-mummies found in votive contexts (those manufactured as part of the votive mummy industry) represent an exciting area of the subject of animal mummification as they are essentially material remains created in antiquity intended to dupe unsuspecting pilgrims. This is not to say that all pseudo-mummies are fraudulent specimens as it is highly probable that a number of them were created from material which the embalmers believed to be sacred in some way, either by association with a sacred animal or a place of worship.

There are potentially a number of reasons as to why pseudo-mummies were created and in order to understand why they may have been made, it is necessary to at first gain an insight into the industry that created them. It can be assumed from the sheer volume of animal mummies at the large necropolises that the production of votive specimens was a lucrative industry bringing huge financial rewards for the temples themselves. Pilgrims are known to have flocked to the cemeteries throughout the year and especially on specific holidays when the sacred animals were worshipped.

The pilgrims were essentially a receptive market through which to sell animal mummies by the hundreds of thousands. The pilgrims believed that through purchasing a mummy, they were able to offer a gift to the god; a gift which acted as an oracular interface between the world of the living and the spiritual world of the gods. They could ask the mummy questions which they believed would be answered by the gods in return in the form of a sign of one kind or another.

The enormous volume of mummies produced indicates that there must have been a well-managed system of procurement, manufacture and point of sale to rival most modern trades. It is the production stage at which the pseudo-mummies were created, and although the reasons behind their production remain speculative, there are a number of possible suggestions.

Firstly, was it simply a question of supply being outweighed by demand? Was the manufacturing process not able to produce enough legitimate animal mummies for the volume of pilgrims? Archaeological evidence from the temples suggests that the animals were bred and raised in captivity within the confines of the complexes. This means that millions of animals would have needed to be housed at any one time which raises issues of logistics in terms of where they were kept, how they were cared for and by whom. Space, manpower and foodstuffs would have all been stretched to the absolute limit to keep the system functioning. It may have been that the number of animals raised in this manner was simply not sufficient to supply the demand for mummies. This theory would be supported by the large number of animal mummies which are found to contain only part of an animal, whether this was a portion of the body or just an isolated bone. By butchering the animals after death it

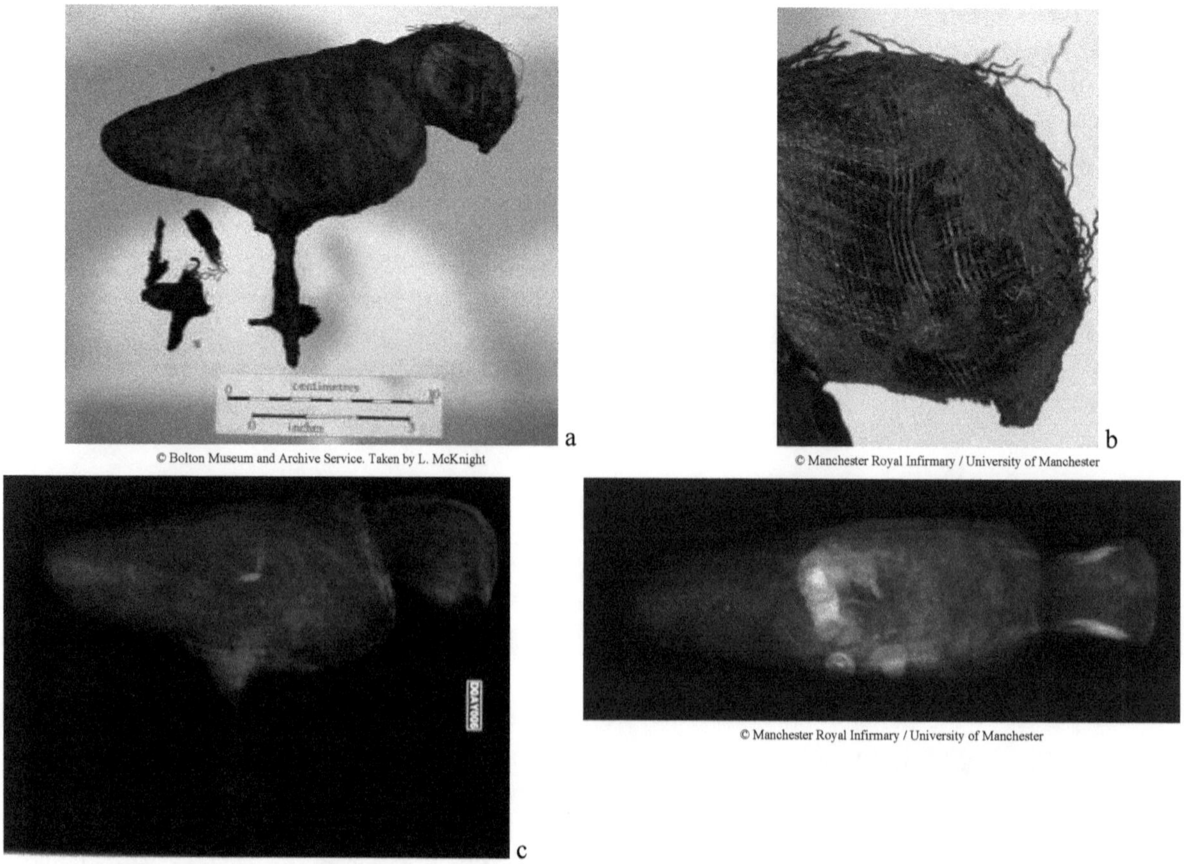

© Bolton Museum and Archive Service. Taken by L. McKnight

© Manchester Royal Infirmary / University of Manchester

© Manchester Royal Infirmary / University of Manchester

© Manchester Royal Infirmary / University of Manchester

Fig. 7.8 – a) and b) Photographs of mummy MA7.19 demonstrating the standing position of the 'bird' and the facial features created by the embalmers, c) AP and d) Lateral radiograph showing that the mummy bundle contains no skeletal material

would have been possible to create multiple mummies from one animal, therefore making the available resources go further.

Secondly, was it a case of the embalmers not being able to cope with the workload thrust upon them as the popularity of animal cults increased? The men employed to slaughter and mummify the animals must have been stretched to their limit during the busiest times of the year. Were they simply not able to work at a fast enough pace? This theory seems unlikely due to the fact that many of the mummies display a great deal of attention to detail indicating that time was taken to produce them carefully. One would assume that if manpower resources were stretched to the maximum then all the resulting mummies would have been of inferior quality and lacking in time-consuming visual adornments.

Thirdly, did the embalmers have selfish reasons for not always manufacturing legitimate animal mummies? It is possible that the embalmers were paid on a piece-work basis, and therefore the more mummies they produced, the bigger the financial reward. This may explain why so many pseudo-mummies were created; however, it raises questions as to why this practice was allowed to take place if it contradicted the doctrines of the temple. Is this an early example of fraudulent behaviour in the workplace?

Fourthly, did it matter that not every supposed animal mummy contained actual remains? Mummies which contained ambiguous material such as feathers, egg shell or fabric, may have been deemed sacred simply through their association with the animals themselves. It may even have been possible that the pilgrims were aware of the fact that not every mummy contained skeletal material and that they purchased mummies on aesthetic value alone. It is likely that the mummies formed part of a pricing structure and it seems realistic to suggest that those that appeared to be more elaborately decorated achieved higher prices. Could it be an issue of wealth and status as to which mummies were purchased and by whom rather than what they actually contained?

It is possible that either one of the four theories discussed above is correct and it is likely that the actual reasons behind the occurrence of pseudo-mummies incorporates elements from all of them. What is certain is that we will probably never know with certainty why pseudo-mummies exist, but they will always present enigmatic and mysterious subjects.

A number of specimens appear so lifelike from their external appearance that doubts can be cast on the nature of their contents even prior to analysis. This was highlighted frequently during the radiographic sessions when it became somewhat of a game amongst the radiographers to second-guess what was inside the bundle as the object entered the scanner! More often than not, those mummies which appeared the most lifelike were in fact found to contain no animal material. It was almost as though the embalmers were simply trying too hard to convince the paying pilgrims that the specimens were legitimate devotional objects; they were simply over-compensating for their lack of diligence to the religious task in hand.

One example which can be seen to demonstrate this theory at its best is that of MA7.19 (Fig. 7.8). This specimen has been modelled to resemble a bird shown in a standing position. Birds were never mummified in this position, but presumably the technique was employed to make the object appear lifelike even down to the individually modelled legs and claws. The head of the mummy shows linen bandaging which has frayed over time and fabric eyes have been added to the external aspect. The specimen almost looks like a child's toy rather than a votive mummy.

The radiographic analysis of MA7.19 shows an altogether different image. The supposed bird mummy is in fact completely void of animal material and appears to have been constructed in its entirety from a mud-like substance bound in linen bandages. The CT images demonstrate that the false eyes have been constructed from small linen fragments with a dense outer layer possibly indicating that a resinous substance was applied to the surface.

Specimens such as MA7.19 are unique examples of the embalmers art at its best. They do not represent skill in the treatment of the physical remains of an animal as they are completely empty of such material. What they do display is an incredible level of attention to detail and artistic licence. Essentially what the embalmers were doing was creating viable objects for sale in a lucrative industry from little more than naturally occurring mud and objects that they found lying around; it really is an example of exceptional business skills if nothing else!

Chapter Eight
OTHER VARIETIES

8.0 – INTRODUCTION

Three specimens studied during the course of the project are in fact neither human nor animal mummies. They were selected as part of the study group as it was believed that they could represent animal mummies despite their unusual appearance. All three specimens display human facial features, yet they seemed to be too small to contain actual human remains. It was at first believed that they may contain isolated human bones or possibly animal remains and as none of them had previously been studied using imaging, it was decided to incorporate them into the study group.

The three specimens are discussed individually below with a discussion section at the end of the chapter.

8.1 – MA8.1

This mummy was described as being 'indeterminate bird or animal mummy – possibly a tourist fake?' in the catalogue of Bristol Museum. The mummy was presented for study in a wooden coffin with a hinged glass lid allowing for easy viewing. Interestingly the wooden sides of the coffin had been covered at some point in decorative paper showing Egyptian motifs.

An external examination of the specimen reveals a conical body shape similar to that given to mummified birds. The human facial features appear clearly on the exterior and are quite stylised in appearance.

Radiography revealed that the mummy bundle contains no human or animal skeletal material (Fig. 8.1). The body shape appears to have been formed from a relatively dense substance which may indicate that resin has been applied during the mummification process, possibly to the bandages prior to wrapping. A transverse fracture is evident through the mid-section of the mummy bundle.

CT highlighted the extent of the damage caused by the fracture which appears to completely transect the mummy. The bandages appear to have been applied in multiple layers in a similar fashion to other mummies within the study group (for example MA7.20). This implies that the mummy was wrapped in three distinct stages and at varying degrees of compression leading to obvious changes in fabric texture on the scan images. CT confirms that there is no skeletal material within the bundle and that there are no visible anomalies as would be expected if there had been unusual foreign objects within. The scans confirm that the bandages appear to have been treated with a radio-dense substance such as resin or tar.

CT computer manipulation software was used to gain a clearer insight into the method of mummification employed in this case. Experimenting with the thresholds and the levels of attenuation of the images made it possible to eliminate the bandages at the base. This revealed the central core of the mummy bundle and the plaster face which appears to have been added as a mask over the mummy.

8.2 – MA8.2

This mummy held in the collection at Bolton Museum and Art Gallery was recorded as being a 'pseudo-baby' due to its unique external appearance (Fig. 8.2). It resembles a human-faced doll with individually modelled limbs and facial features of dark brown colour. One leg has been shortened by the loss of the end of the limb and the stomach shows an exaggerated concave appearance.

The radiographic analysis of this specimen revealed that it contains no human or animal skeletal material and seems to have been formed from a mud-like substance. Interestingly, a wooden stick can be seen protruding from the skull area down into the chest presumably as a support around which the body shape was formed.

The mummy displays human facial features added in the form of a mask. Air bubbles are visible within the

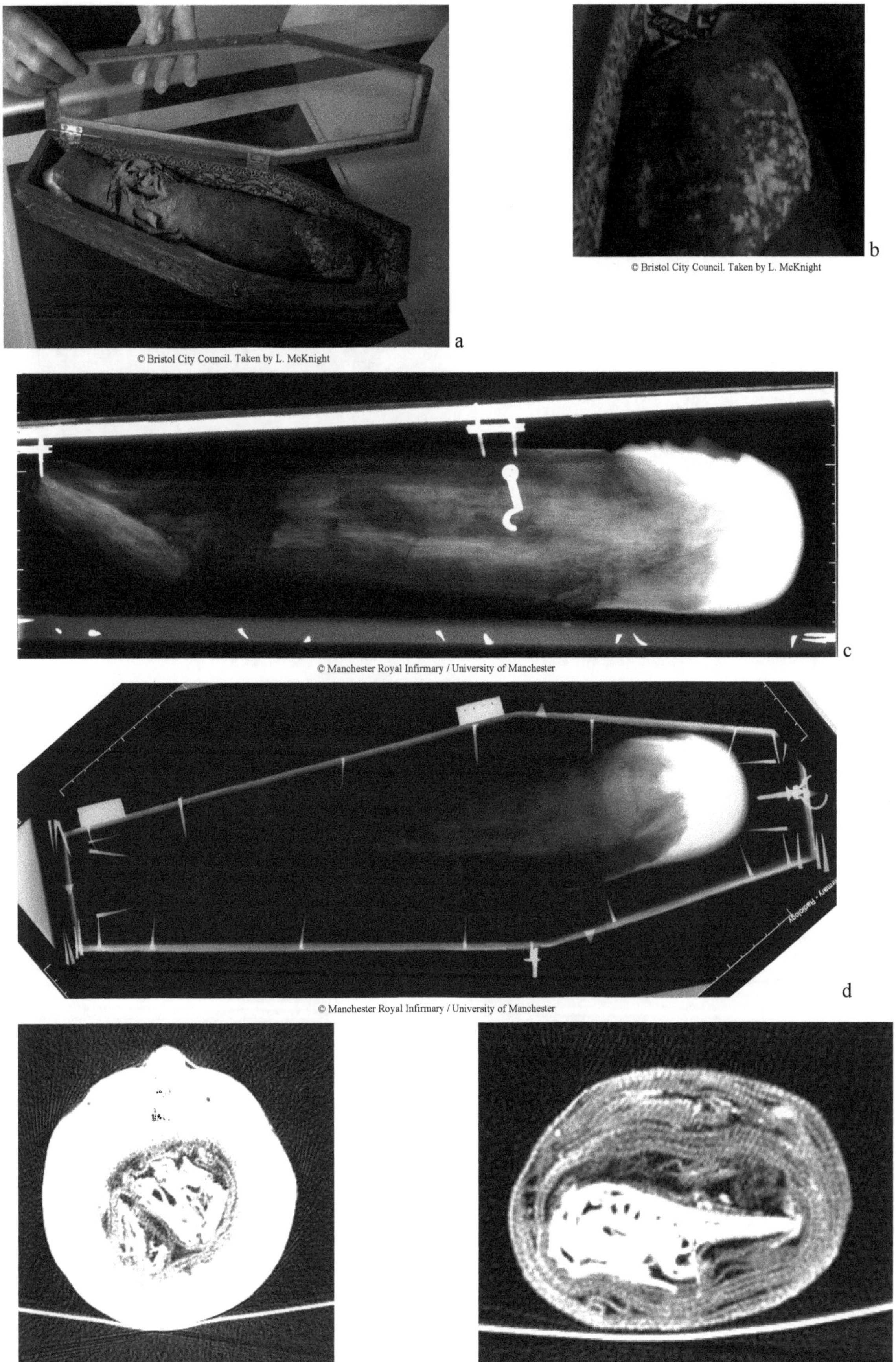

© Bristol City Council. Taken by L. McKnight

© Bristol City Council. Taken by L. McKnight

© Manchester Royal Infirmary / University of Manchester

© Manchester Royal Infirmary / University of Manchester

© Manchester Royal Infirmary / University of Manchester

© Manchester Royal Infirmary / University of Manchester

Fig. 8.1 – a) and b) Photographs of mummy MA8.1 showing the glass lidded coffin and the stylised plaster face, c) Lateral radiograph, and d) AP radiograph showing the radio-dense plaster mask over the head and face and the complete absence of skeletal material. Images e) and f) CT axial sections showing the use of rolled bandages to create the mummy shape and a section through the plaster mask

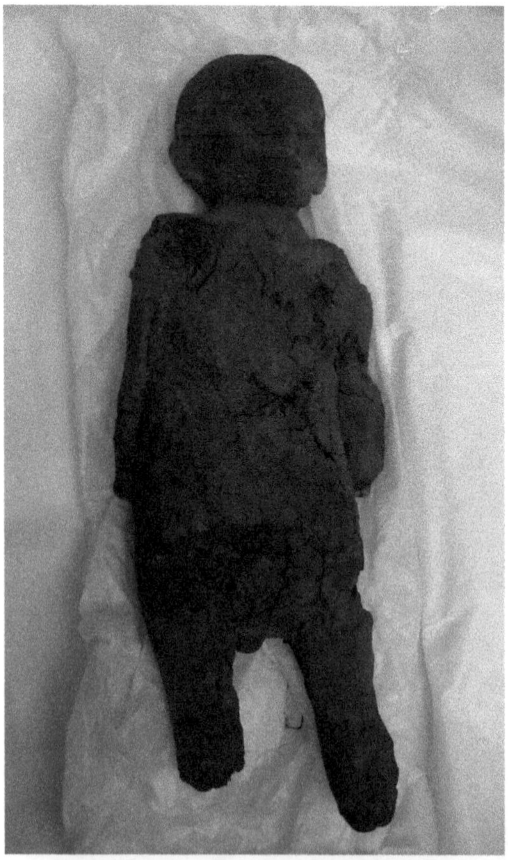

a

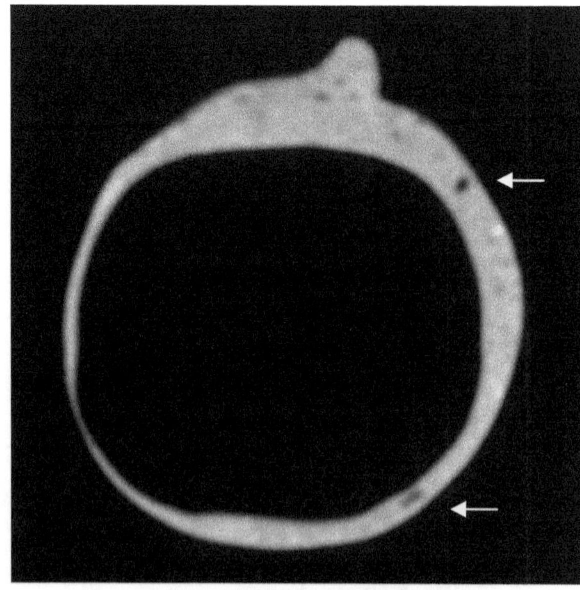

b

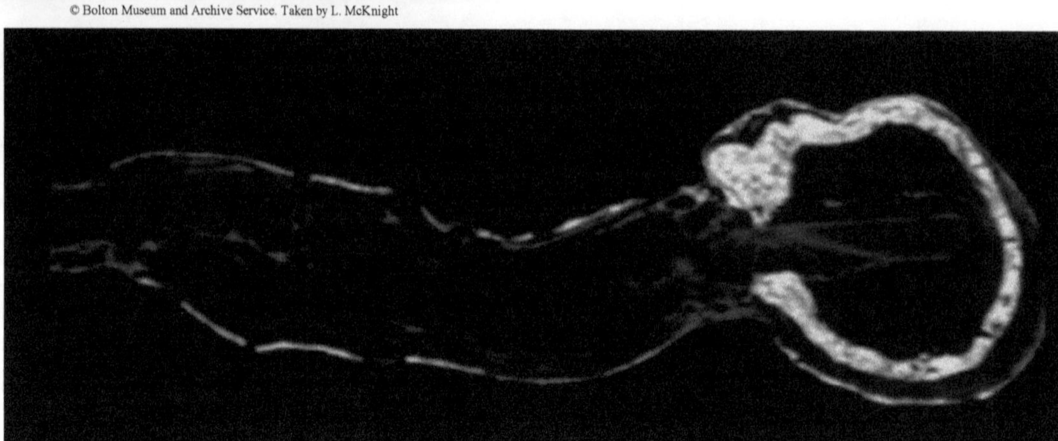

c

Fig. 8.2 – a) Photograph b) Axial CT slice through the facial mask on mummy MA8.2 showing the presence of air bubbles within the plaster mixture, c) CT reformation of mummy MA8.2 demonstrating the concave nature of the modelled body and the presence of what appears to be a wooden stick inserted through the neck to provide support. The profile of the plaster face is clearly visible

substance used to create the mask which resembles Plaster of Paris.

8.3 – MA8.3

This specimen is also from the collection held at Bolton Museum and Art Gallery and it shares a number of similarities with specimen MA8.1 (Fig. 8.3) – it displays the conical body shape given to mummified birds, it has human facial features, and it is presented in a wooden coffin with a glass lid.

Radiography again showed that this mummy contained no human or animal material. The body shape appears to have been created from a mud-like substance wrapped in layers of bandages. The face appears to be in the form of a mask made from a substance similar to Plaster of Paris.

8.4 – DISCUSSION

Radiographic analysis of these three unusual objects revealed that they are in fact neither human nor animal in origin. They appear to all have been created from organic

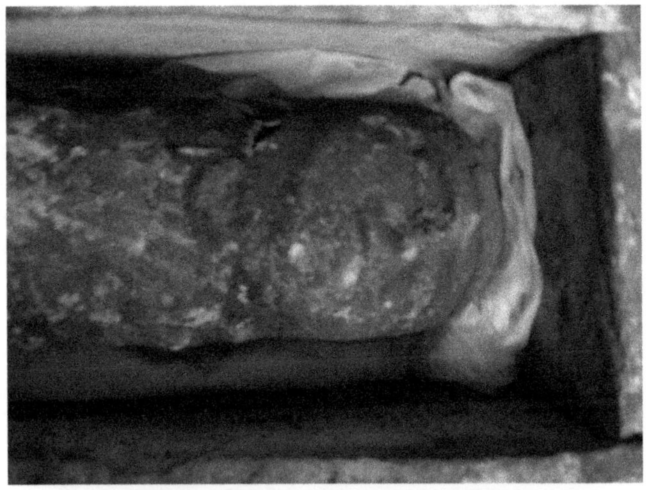

a

© Bolton Museum and Archive Service. Taken by L. McKnight

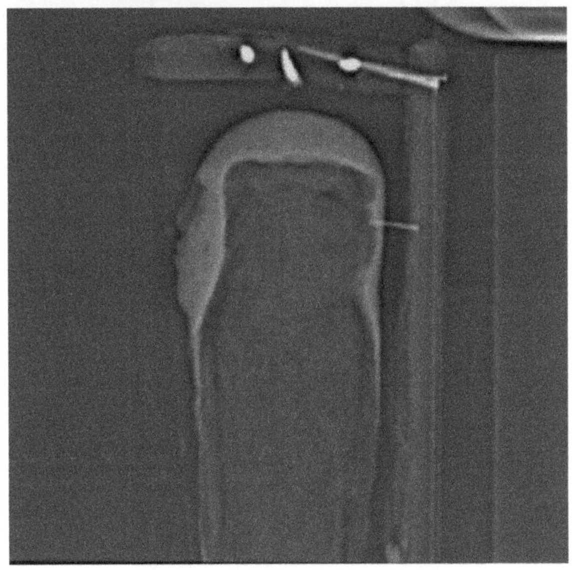

b

© Manchester Royal Infirmary / University of Manchester

Fig. 8.3 – a) Photograph and b) Scout projection image of mummy MA8.3 created during the CT process. The image clearly shows the plaster facial mask which has been applied over the head of the mummy bundle. Again the human facial features appear to be stylised

materials before being given plaster masks displaying human facial features. As we have little provenance or dating information for these specimens it is difficult to assess which time period they can be attributed to with any certainty.

In the case of MA8.1 and MA8.3 I would like to suggest that the mummy bundles themselves are in fact ancient pseudo-mummies possibly originally manufactured to resemble bird mummies. This is because the wrapping techniques demonstrated in these two cases suggest that the objects are genuine ancient Egyptian specimens due to the levels of similarity between them and other specimens within this study group. In particular the three distinct layers of bandages displayed in the case of MA8.1 reinforces this assumption.

Mummy MA8.2 is certainly the exception amongst these three peculiar objects as it displays few similarities to the other two except for the human facial features. By all accounts it resembles a doll, a small object made in human form for reasons unknown. Due to a lack of provenance it has been impossible to suggest a date for the specimen. It could be an ancient Egyptian object intended for purposes unknown or it could be a modern forgery purporting to be ancient. Further biomedical analysis may help to increase our understanding of this object in the future.

The masks displaying human facial features shown in all three cases are unusual and slightly disturbing to look at as they are tiny, but accurately formed, stylised faces. It was deemed to be outside of the scope of this study to attempt to acquire samples from these specimens for biomedical analysis, but it is hoped that in the future this work may be attempted to assess the substance from which the masks have been created. From a radiographic point of view, it seems reasonable to suggest a substance similar to Plaster of Paris.

It is feasible that specimens MA8.1 and MA8.3 are in fact ancient mummies but that they have been modified to incorporate the human facial mask and wooden presentation coffin at some point in more recent times, possibly the Victorian era when the fascination with ancient curios reached a peak. The origins of MA8.2 are more unclear; it is either an ancient mummy bundle modified later, or it is a completely modern specimen.

Speculating that these specimens are all unique and rather odd trinkets featuring modern modifications can only be explained if they were created to supply a demand for souvenirs of some kind. The style of presentation of MA8.1 and MA8.3 in glass-lidded coffins adds to the speculation that they may have contained human remains; however, this study can at least clarify that this is simply not true. What can be said is that the mystery surrounding these objects remains unexplained.

Chapter Nine
CASE STUDIES

9.0 – INTRODUCTION

In order for the reader to gain a more comprehensive insight into the nature of the mummified material studied during the course of this research project, the decision was made to include a chapter solely based on case studies. These should be used to supplement the information provided in Chapters Five to Eight and can be used in conjunction with the complete catalogue which is included in the Appendix of this study.

The case studies described below have been chosen from the study collection as they help to illustrate either the potential value of radiography or the problems associated with its application to mummified animal material. The italicised writing below each heading relates to the information provided by the museum records (if available). The abbreviation XR refers to information gained through study using radiography and CT refers to computed tomography.

Reproductions of the radiographs have been included alongside photographs of the specimens where available to provide the reader with a true sense of each specimen under discussion and to show evidence of particular points being made.

9.1 – BIRDS

MA1.1

'finely wrapped in square design with false foot. Late Period? Egypt'

Identification – cf. Falcon. Nostrils similar to *Accipiter nisus* (Sparrowhawk)

Radiographic Report – XR – The mummy bundle appears to be fractured (separated) in the region of the tail. Interestingly, the skeleton itself is much smaller than the mummy bundle with the skull having been placed within the modelled body, not in the area of the false head. The false head appears to have a radio-dense coating which could be resin or gilding. There appears to be an area of soft-tissue or feathers which is extraneous to the bird, but is included within the bundle wrappings. This could be the tail of another bird.

CT – There is clear evidence to suggest that there is a thin layer (possibly even a single layer) of bandages around the skeleton of the bird which have a radio-dense appearance possibly indicating the application of a layer of a substance such as resin, paint or natron. The bundle has then been wrapped again to form a more uniform cylindrical shape with what appears to be rolls of bandaging on each side of the neck which were presumably added for support. The bundle has then been rewrapped with the inclusion of particles of a denser material (possibly wax) laid intentionally in a row within the wrappings. The mummy has been given a false head and feet.

Conclusions – CT has been very informative in this case providing a clear insight into how the wrappings have been applied by the embalmers. The CT images are beneficial in providing information on the dense particles thought to be wax which have been laid intentionally in a row within the third layer of bandages.

MA1.28

Identification – *Falco naumanni* (Lesser Kestrel)

Radiographic Report – XR – This specimen consists of a complete mummy bundle containing the skeleton of a bird. The distal half of a juvenile crocodile skeleton can be clearly seen positioned alongside the breast of the bird. The skeleton of the bird displays evidence suggestive of a pathological defect to the spinal column of the bird and what appears to be a compression to the vault of the skull. There is an area of density located in the region of the neck and the tail, and a mass is visible in the gut. The outer layer of bandages appears to be loose.

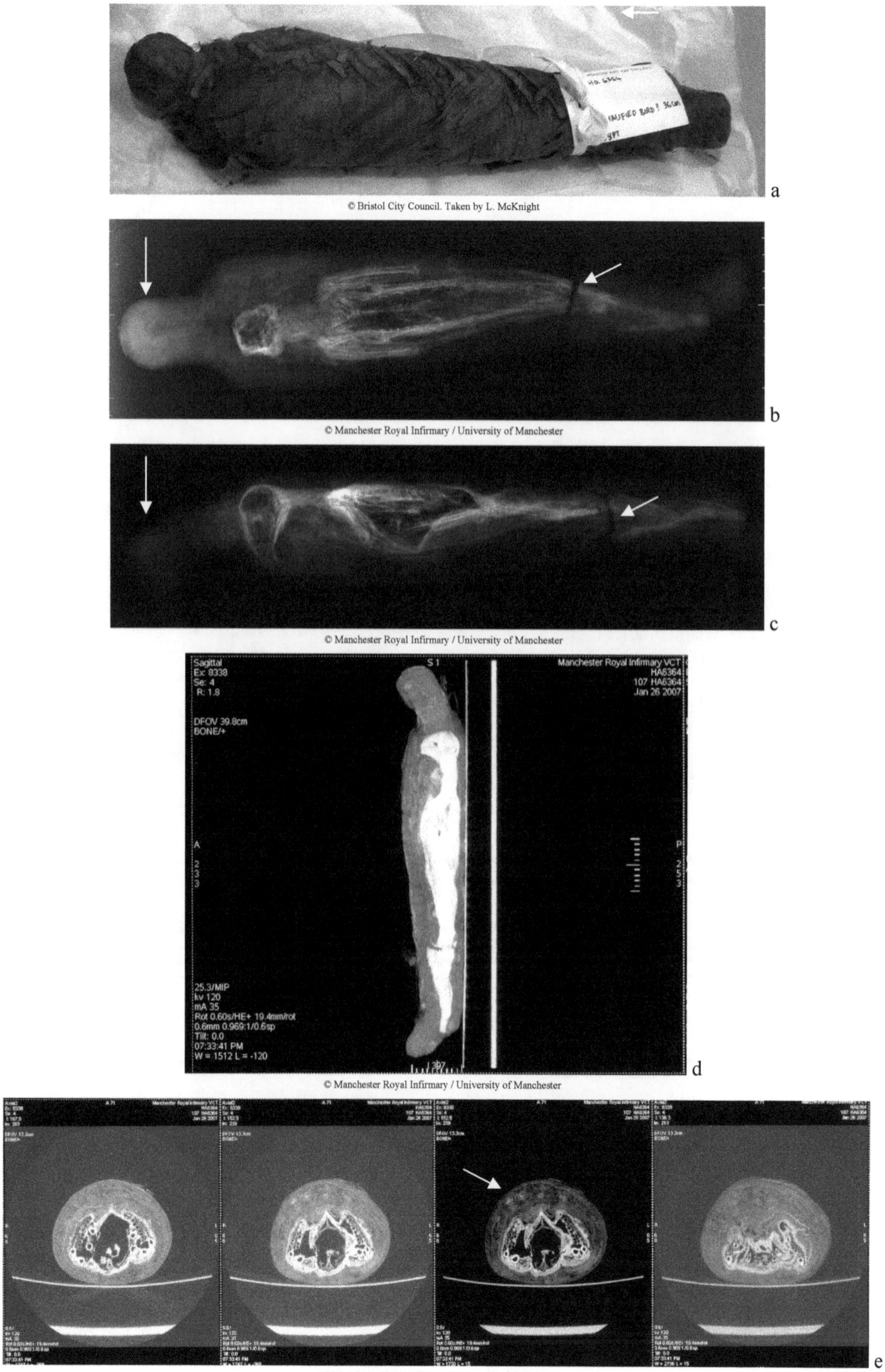

Fig. 9.1.1 – a) Photograph, b) AP radiograph, c) Lateral radiograph, d) CT reformation and e) CT axial sections of mummy MA1.1 showing the post-mummification fracture through the tail. The skeleton does not fill the bundle and a false head has been added. Image e) shows the presence of a row of what appear to be intentional inclusions of a wax-like substance which are not visible on the X-rays. It is also possible to view three distinct layers of bandaging that have been applied to the bird during the mummification process

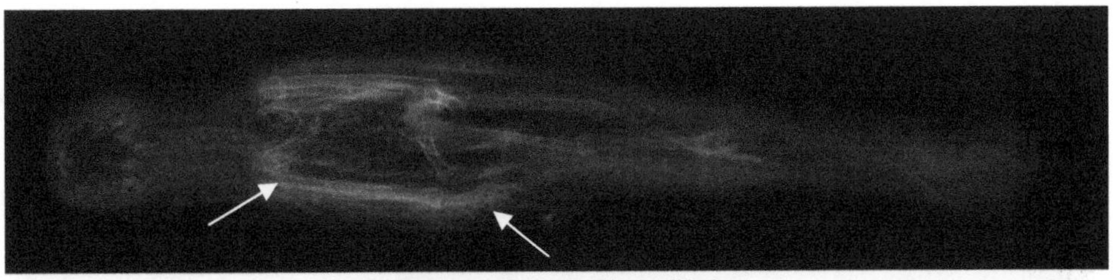

a

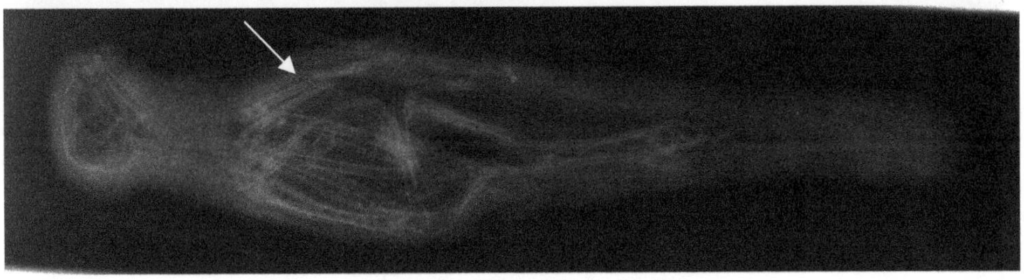

b

Fig. 9.1.2 – a) AP radiograph and b) Lateral radiograph of mummy MA1.28. Both of the radiographs highlight the presence of the distal half of a juvenile crocodile skeleton placed alongside the breast. Image a) provides evidence of what could be a pathological defect to the spinal column of the bird causing curvature of the thoracic region

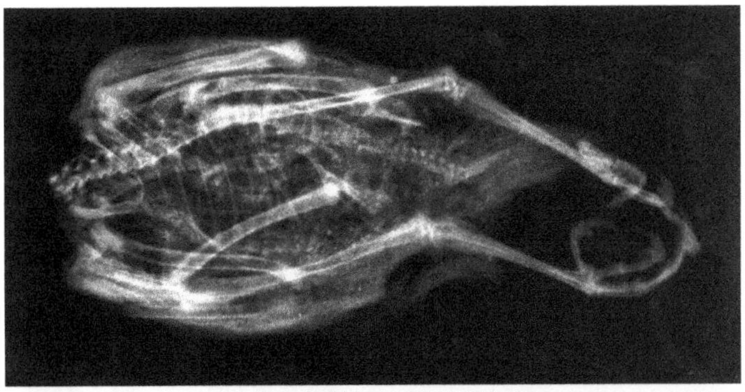

a

b

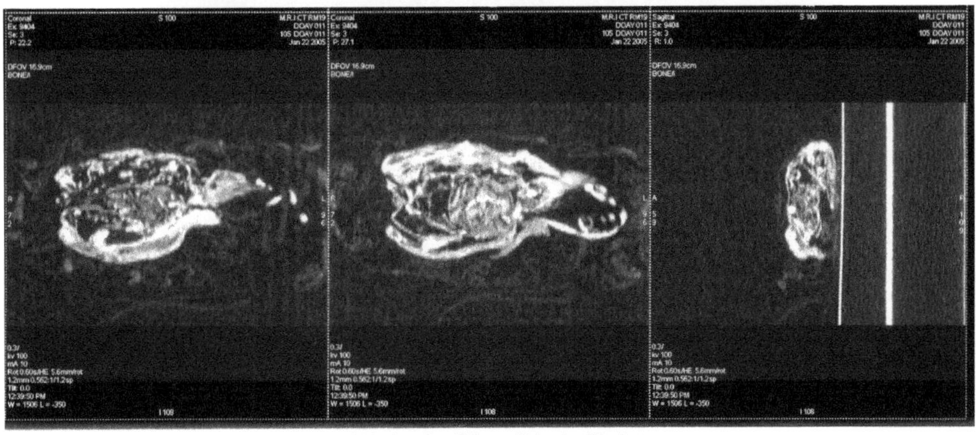

c

Fig. 9.1.3 – a) AP radiograph, b) Lat radiograph and c) CT axial sections of mummy MA1.32. This mummy demonstrated excellent spatial resolution on the radiographs, which was found to be reduced on the CT images. The bird is wrapped only in a thin layer of bandages with the claws exposed. Image c) shows the presence of a mass in the abdominal cavity

CT – Experiments were carried out with the threshold of the images in an attempt to gain a better insight into the bundle contents. These experiments revealed that altering the attenuation using computer software can be used to visualise the remaining soft-tissue which has been preserved through the mummification process. This method was also used to show the crocodile skeleton more clearly. As the images were magnified to look at particular areas of interest it was noted that detail was reduced. This is shown well when the images are converted into reformations where the CT data is reconfigured to view the animal in three dimensions.

Overall, this specimen provides an excellent case study for the advantages and disadvantages of CT as a research technique.

MA1.32

Identification – *Falco tinnunculus* (Kestrel)

Radiographic Report – XR – This mummy bundle contains the partial skeleton of a mature bird. The skull is not present and one wing is fractured. There is a mass in the gut area which appears to contain small bones which is thought to represent organ residue or stomach contents. The bandages appear on the images to consist only of a thin layer and the claws are uncovered and can be seen protruding from the wrappings.

CT – This specimen demonstrates the poor quality spatial resolution which CT provides compared to X-ray.

Overall, CT appears to add little in the way of further information to supplement that gathered through X-ray. This is one specimen which revealed clearer and more informative images through X-ray alone. The application of CT to the mummy bundle highlighted the issue of reduced spatial resolution which occurs as a result of the loss of data from the areas located between the axial sections.

MA1.35

Identification – *Falco tinnunculus* (Kestrel)

Radiographic Report – XR – The bundle is incomplete as the skull has become detached as a result of its desiccated condition. The bundle was found to contain only two bird legs complete with claws. There is a detached skull under the same museum acquisition number included with this mummy bundle although the proportions of the skull indicate that it is unlikely to belong to the same animal as the legs within the bundle. The bandaging appears to be thick in places and has been moulded into the cylindrical shape characteristic of a bird mummy.

CT – The CT images highlight the presence of pinion feathers that have been used to reinforce the shape of the mummy bundle. In the absence of a complete skeleton, the embalmers would have relied on other materials to provide the body shape.

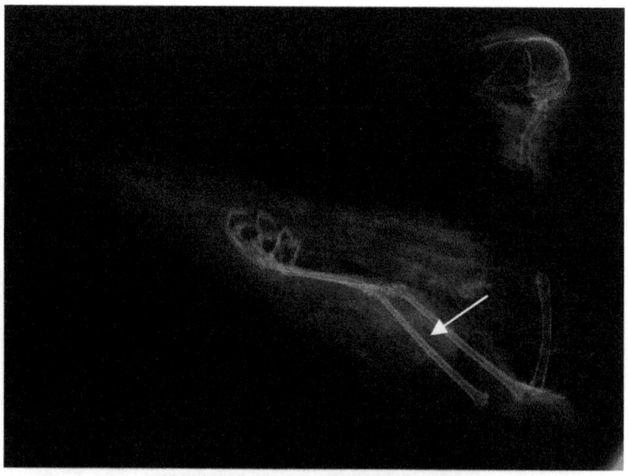

Fig. 9.1.4 – Radiograph of mummy MA1.35 which includes a decapitated skull thought to belong to a different bird. The linear structures forming the bundle are thought to be pinion feathers used to elongate the shape

9.2 – CATS

MA2.1

'bandaged with facial features painted on. Egypt'

Identification – feline, juvenile

Radiographic Report – XR – This bundle contained the largely disarticulated skeleton of a juvenile cat. The skull vault appears to have collapsed inwards leaving a mass of bone fragments at the rear of the cranial vault. The neck has been broken post-mortem possibly due to the restrictive plastic strap positioned around the kitten's neck which has been applied at some point during its storage or conservation to keep it fixed firmly to its presentation board. There appears to be a small radio-dense inclusion within the bundle which could represent a metallic flake, sand or a stone particle.

CT – The images clearly show the false ears and eyes which have been applied to the mummy bundle. There is evidence to suggest that the bundle has been wrapped twice, but no radio-dense coating appears to have been applied.

MA2.2

'Carefully wrapped in different coloured bandages, arranged in a striped pattern. Facial features painted on. Beni Hasan. Mrs A. E. Hudd'

Identification – Feline, juvenile

Radiographic Report – XR – The skeleton is intact and fully articulated and appears to fill the mummy bundle completely. One ear protrudes from the bandages and there is a dense area on it which is thought to be resin applied during the mummification process. There are a

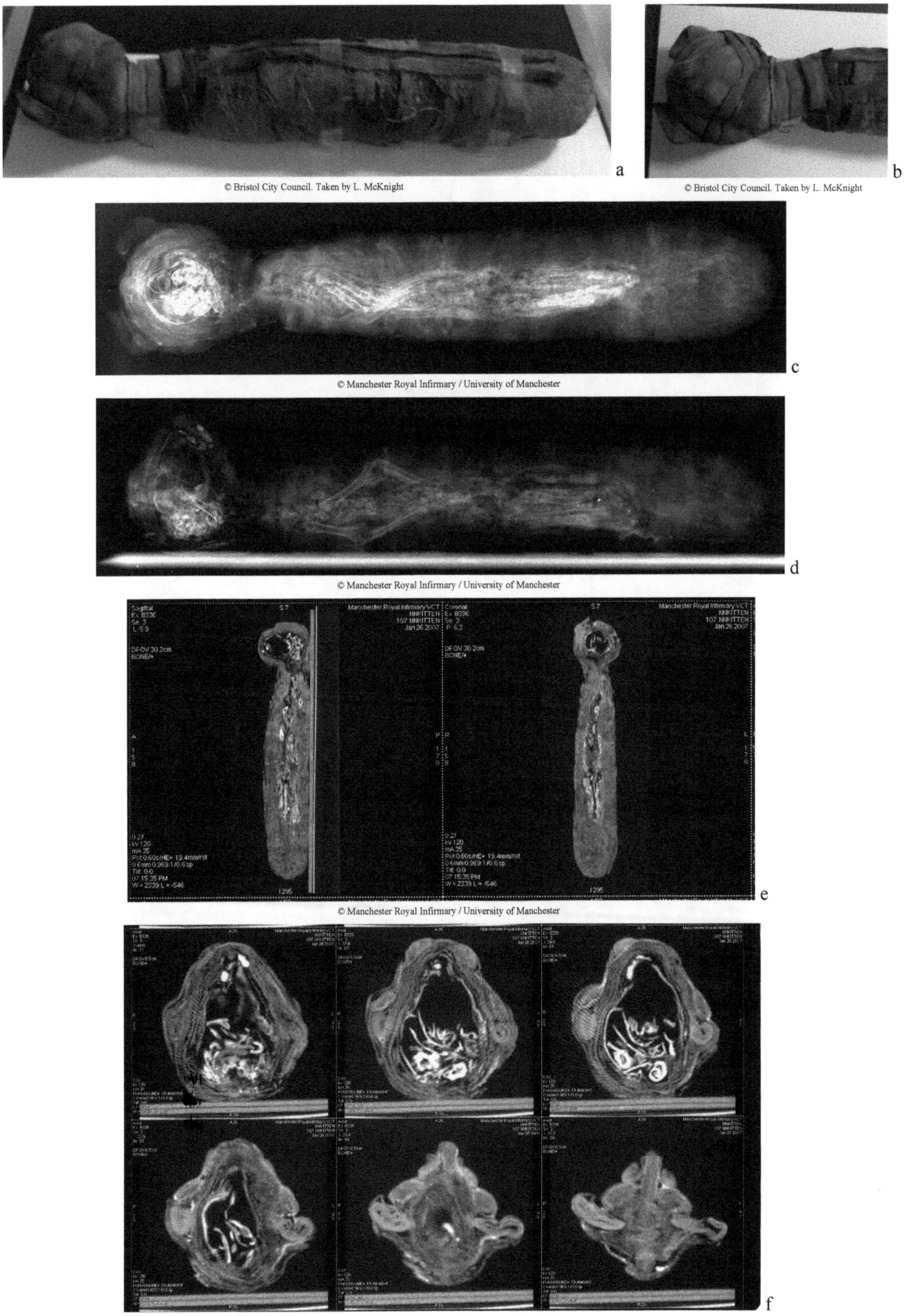

Fig. 9.2.1 – a) and b) Photographs, c) AP radiograph, d) Lateral radiograph, e) CT reformations, and f) CT axial sections of mummy MA2.1. The images show the presence of false ears and the collapse of the skull vault leaving a mass of bone fragments

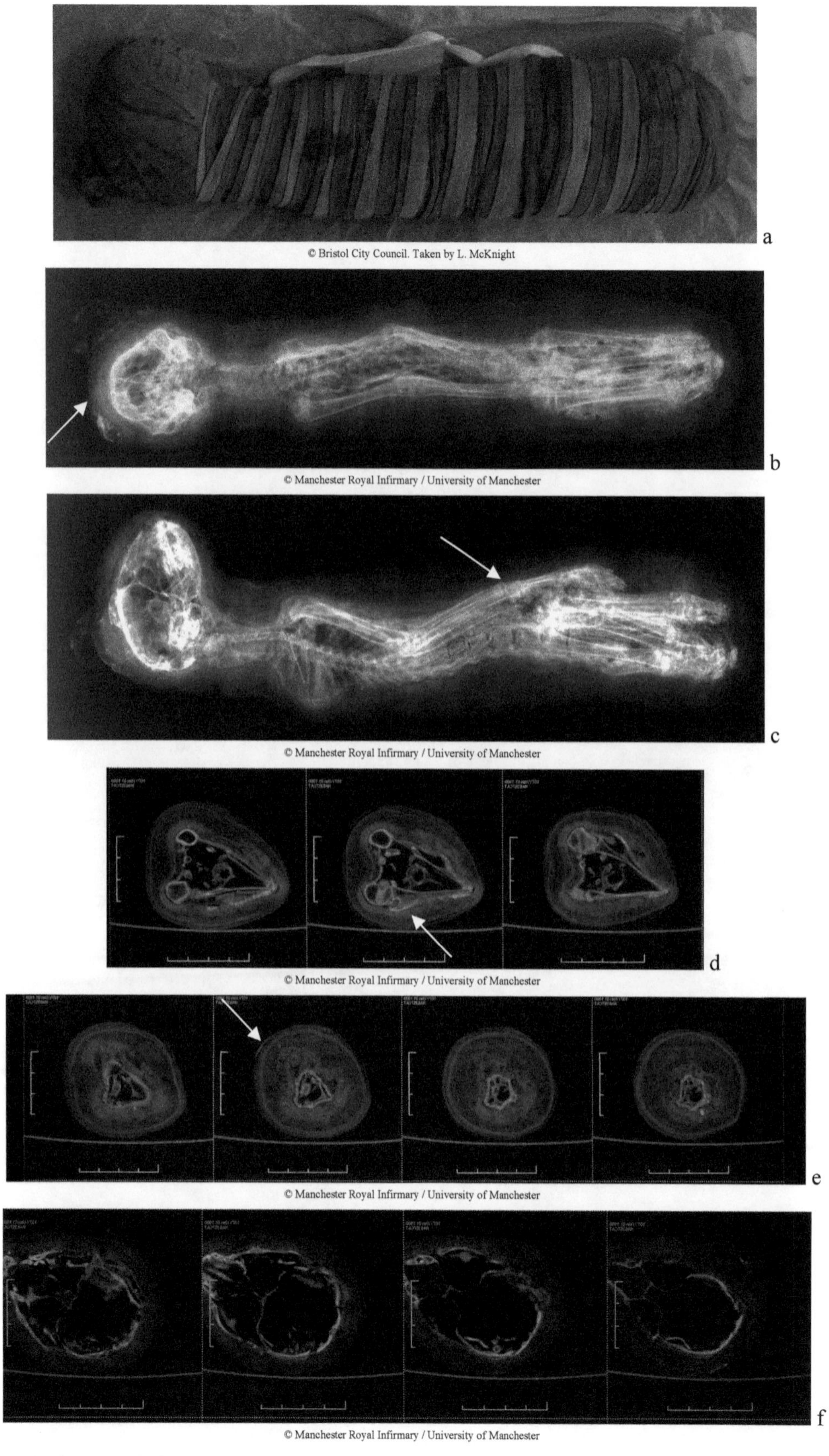

Fig. 9.2.2 – a) Photograph, b) AP radiograph, c) Lateral radiograph, d), e) and f) CT axial slices. Image b) shows the presence of a radio-dense area on the ear which protrudes from the wrappings which is thought to indicate resin used during the mummification process. Image c) highlights unfused epiphyses on the limb bones indicating that the animal was juvenile at the time of death. Image d) nicely illustrates the blades of the scapuli. Image e) demonstrates continuity of the spinal column suggesting that there are no fractures to the neck, and it is also possible to clearly see the three distinct layers of bandaging on the mummy. Image f) shows that the orbital sockets are free of dried matter and false inclusions

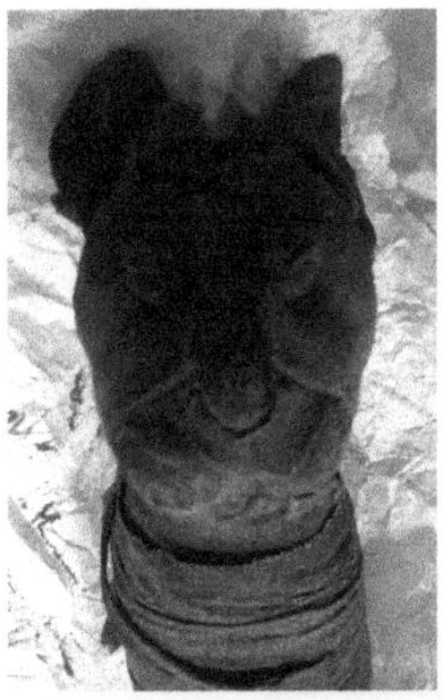

a

© Bristol City Council. Taken by L. McKnight

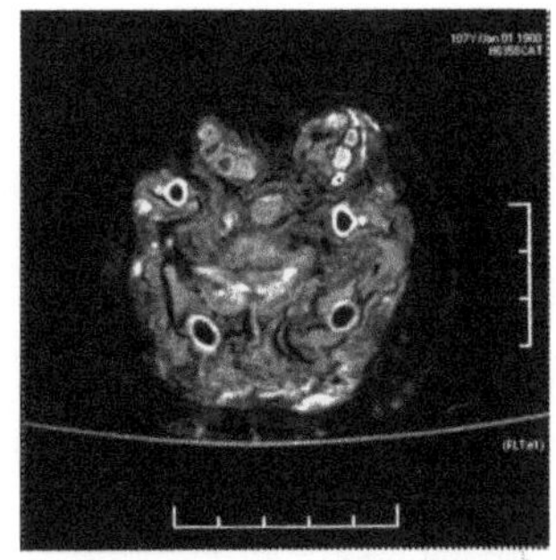

d

© Manchester Royal Infirmary / University of Manchester

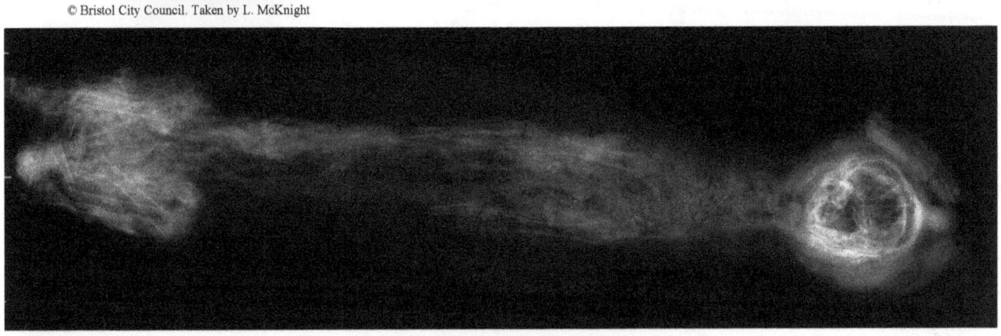

b

© Manchester Royal Infirmary / University of Manchester

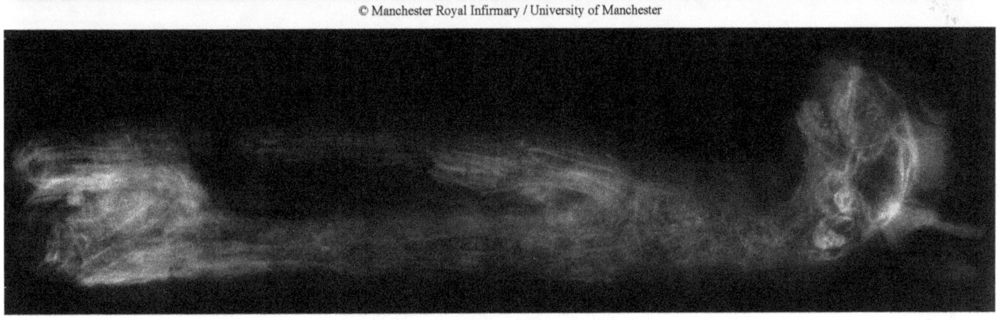

c

© Manchester Royal Infirmary / University of Manchester

Fig. 9.2.3 – a) Photograph, b) AP radiograph, c) Lat radiograph and d) CT axial section. Images b) and c) clearly demonstrate the extensive damage to the skeletal integrity of the animal. Image d) shows the rolls of fabric used to support the fragile skeleton

number of unfused epiphyses visible and the secondary dentition is present high in the tooth sockets suggesting that the animal was immature at the time of death. The hind legs are flexed upwards in the 'normal' manner in front of the chest.

CT – The images show that the orbital sockets are completely empty with no evidence for false eyes or dried orbital matter. There appear to be three distinct layers of bandages surrounding the skeleton. CT also clearly shows the wings of the scapuli and a nicely articulated skeleton.

MA2.3

'Roughly bandaged, though the face is carefully modelled from bandages with 'buttons' of linen for eyes. Egypt'

Identification – Feline

Radiographic Report – XR – The skeleton shows extensive disruption to the skull with the displacement of the petrous bones towards the rear of the skull cavity. There is also a loss of skeletal integrity in the vertebral column. The thorax has collapsed inwards and some ribs appear to have been displaced to the neck region. The pelvis is not visible although there are some unidentifiable bony fragments. One forelimb is fractured, the other is intact. This mummy could represent a partially disarticulated skeleton that has been remodelled to appear more like a cat.

CT – The CT images shows evidence of rolls of fabric within the mummy bundle which may have been added by the embalmers to provide support to the fragile skeletal remains.

MA2.5

'Roughly bandaged. Egypt'

Identification – Feline, juvenile

Radiographic Report – XR – The bundle has been carefully modelled to resemble a complete cat mummy,

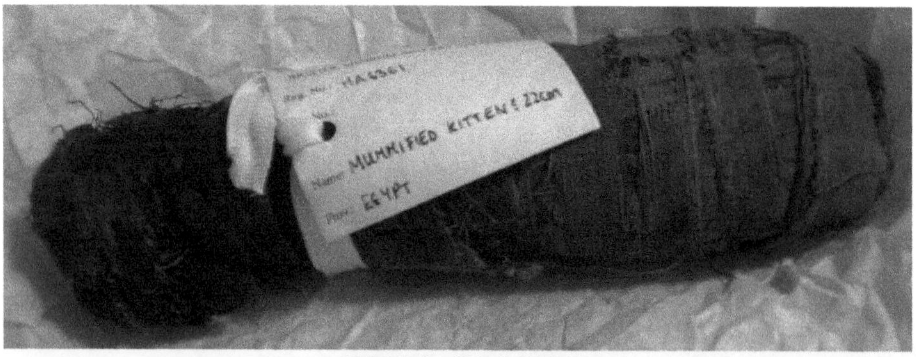

a

© Bristol City Council. Taken by L. McKnight

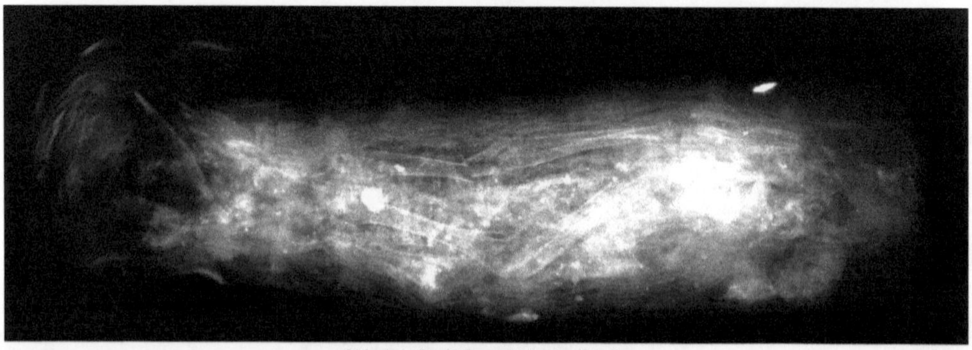
b

© Manchester Royal Infirmary / University of Manchester

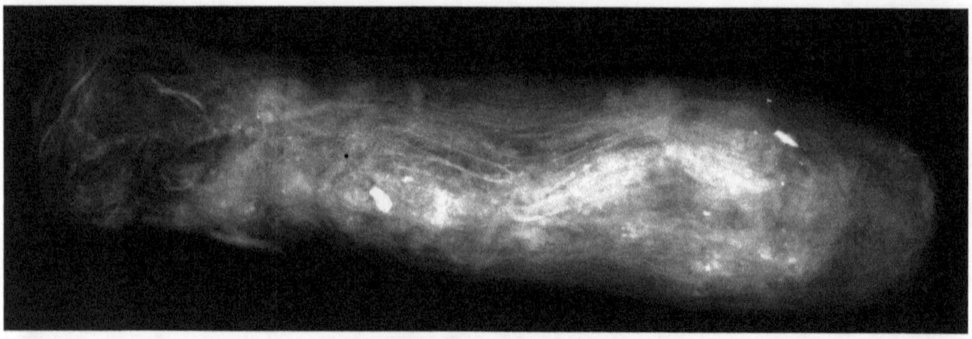
c

© Manchester Royal Infirmary / University of Manchester

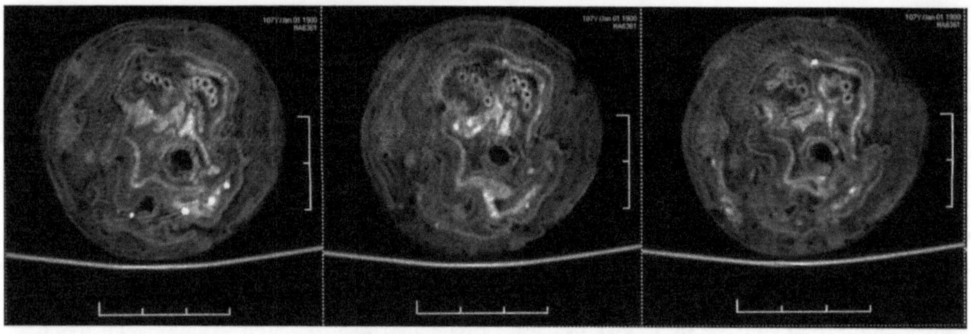
d

© Manchester Royal Infirmary / University of Manchester

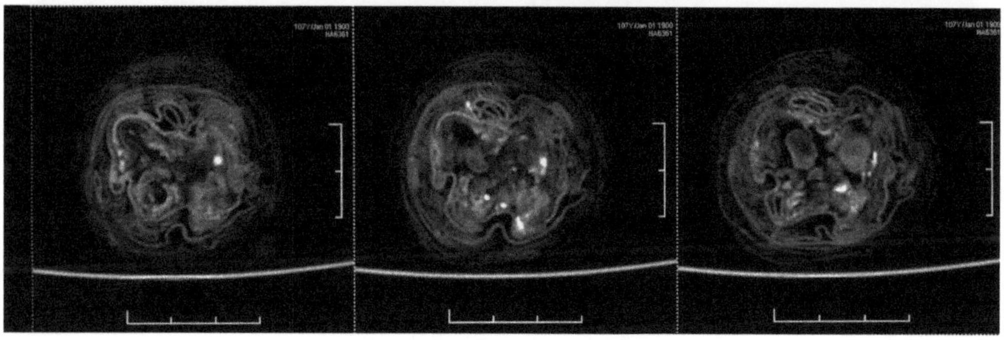
e

© Manchester Royal Infirmary / University of Manchester

Fig. 9.2.4 – a) Photograph, b) AP radiograph, c) Lat radiograph, d) and e) CT axial sections. Images b) and c) show the presence of a partial skeleton with granular inclusions. Images d) and e) are sections through the bundle showing the presence of granular fragments and different layers of bandages. Image d) shows a sectional view of the paws of the cat

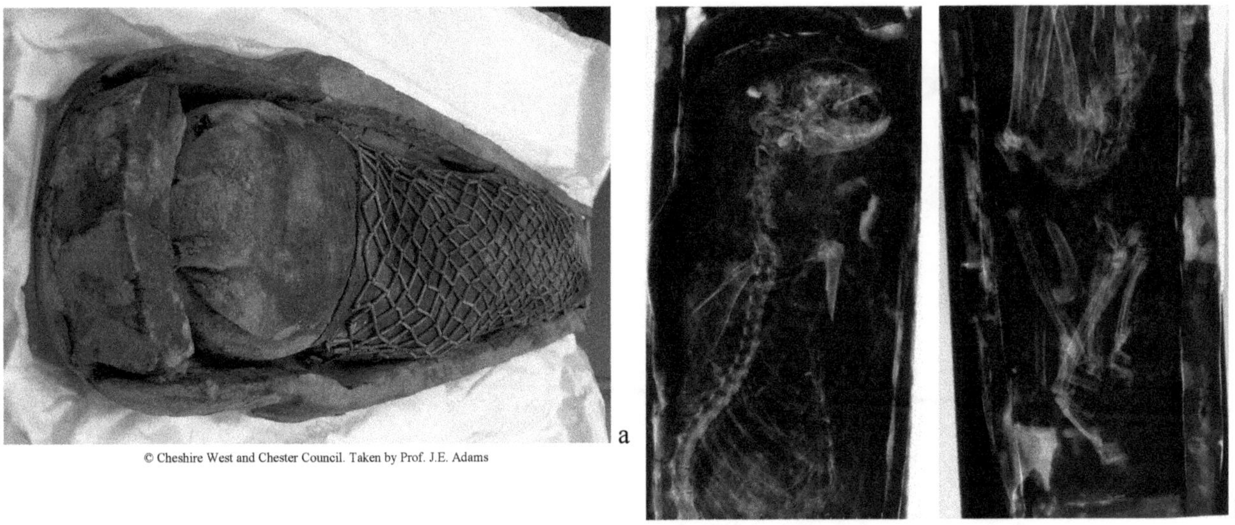

© Cheshire West and Chester Council. Taken by Prof. J.E. Adams

© Manchester Royal Infirmary / University of Manchester

Fig. 9.2.5 – a) Photograph and b) AP radiographs of mummy MA2.6 showing the extensive loss of skeletal integrity within the mummy bundle

but the radiographs show that only two front limbs are present. There is no skull visible. The humerus displays a post-mortem fracture. The wrappings appear to be loose at the 'head-end' and there is a transverse fracture of the mummy bundle in the neck region.

CT – The CT shows that this partial cat skeleton has been wrapped once, dipped in a radio-dense substance then wrapped again. It appears as if the body of the cat could have been formed from an unknown material (possibly sand or mud), with the bones of the front limbs being placed alongside in their correct anatomical position.

MA2.6

Chester Mummy

'Small anthropoid mummy case containing a large cat with some additional bones. Late period. X-rayed by Professor Harrison of the Dept. of Anatomy in March 1973. Given to Grosvenor Museum, Chester, in 1888 by Rear Admiral Sir Richard Massie Blomfield. On long term loan to Liverpool since February 1974.' This mummy is currently on display in the Garstang Museum, University of Liverpool.

Identification – Feline

Radiographic Report – XR – The large size of this mummy bundle dictated that it was imaged in sections with areas of overlap to avoid any omissions. It was also necessary to radiograph the mummy within its wooden coffin although it was possible to remove the lid.

It is a complete mummy bundle containing the skeletal remains of one cat. The majority of the exterior of the mummy bundle is covered with a beaded layer which appears as a mesh on the images. There is extensive evidence to suggest that the skeleton has been dismembered post-mortem. The lower portion of the mummy bundle contains the remains of the lower limbs, partially overlapping.

This specimen highlights the issue of the magnification of the bones which are lying at greater distances from the film and the problems associated with the superimposition of structures. The resulting images provided an excellent preliminary insight into the contents of the bundle which were then elaborated further using CT.

CT – The integrity of the skeleton has been jeopardised leading to extensive displacement of bones within the bundle; however it is believed that the complete skeleton is present. There is a clean fracture of the humerus caused during the intentional post-mortem dismemberment. The limb bones show the presence of visible epiphyses which suggests that the cat was not fully mature at the time of death. The tail appears to have been severed and has been positioned at the base of the mummy bundle.

Conclusions – Mummy MA2.6 was a unique specimen to study for a number of reasons. It is thought to be a high-status animal, possibly even a pet, although without any supporting documentary evidence it has not been possible to ascertain for certain why this mummy was created and the exact status of the animal. The method of presentation and wrapping indicate that great care and attention to detail was taken by the embalmers during the mummification process and the beaded decoration indicates that it could well represent a high-status animal.

This specimen demonstrated the problems caused by superimposition and the magnification of bones on conventional radiographs. The beaded mesh on the exterior of the mummy bundle is clearly visible on the images and it is possible that this is masking features underneath. CT scanning helped to eliminate these issues as it was possible to view the bundle in axial sections.

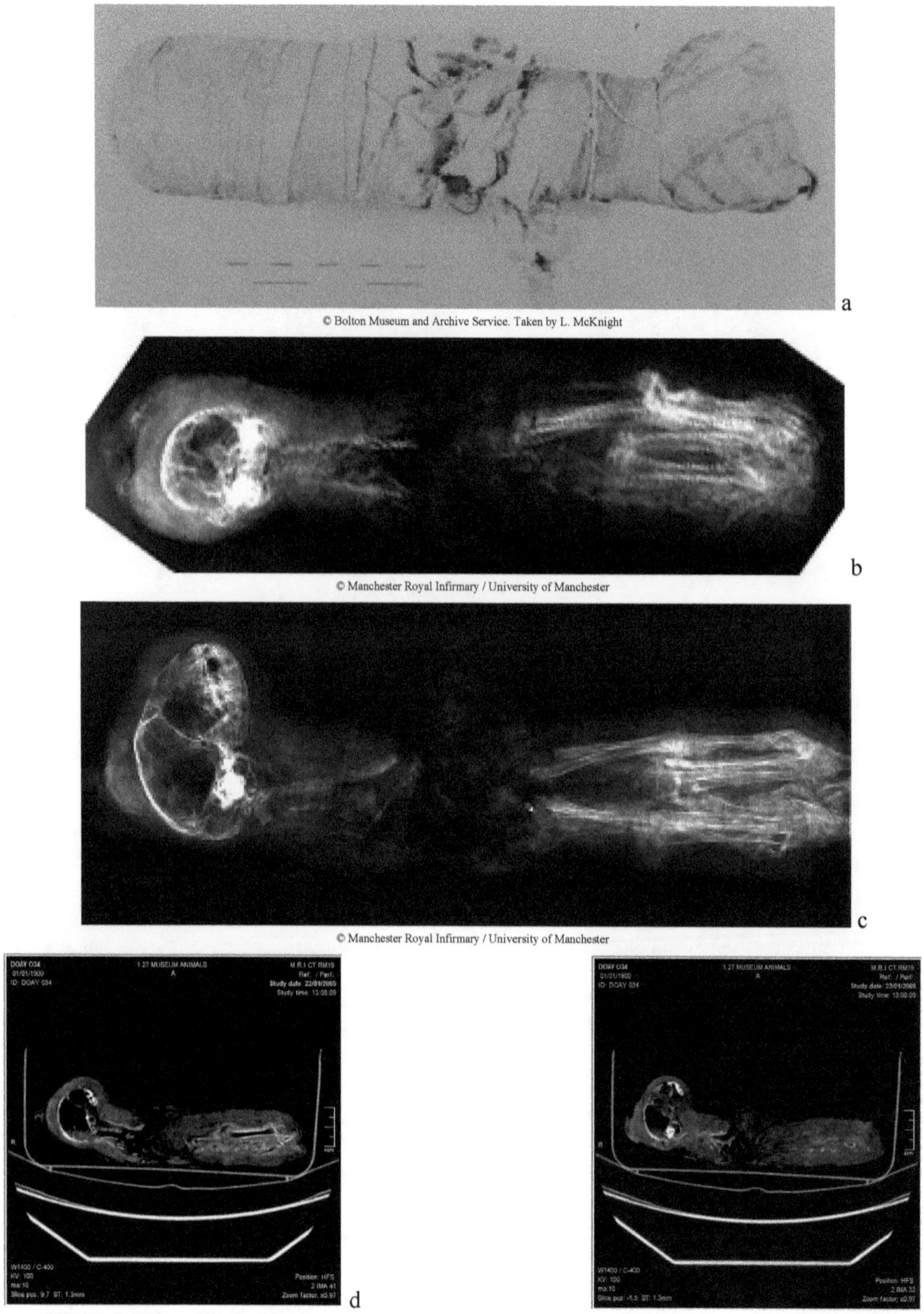

© Bolton Museum and Archive Service. Taken by L. McKnight

© Manchester Royal Infirmary / University of Manchester

© Manchester Royal Infirmary / University of Manchester

© Manchester Royal Infirmary / University of Manchester

© Manchester Royal Infirmary / University of Manchester

Fig. 9.2.6 – a) Photograph, b) AP radiograph, c) Lat radiograph, d) and e) CT axial sections showing the extensive damage to the mid-section of the mummy bundle. Images d) and e) show the presence of dehydrated brain matter at the rear of the cranial vault

MA2.18

Identification – Feline, juvenile

Radiographic Report – XR – This specimen is an incomplete mummy bundle containing the skeleton of a juvenile cat. There is a fracture in the spinal column leading to the separation and loss of vertebrae and a portion of the thorax. The body has been densely wrapped with only the bandages holding the two halves of the skeleton together. The exterior of the bundle has been lavishly decorated with the addition of painted facial features. Brain matter is visible in the skull cavity. Interestingly, the tail appears to have been removed post-mortem, but prior to mummification.

CT – The CT study of this specimen confirms the absence of certain skeletal components; however the

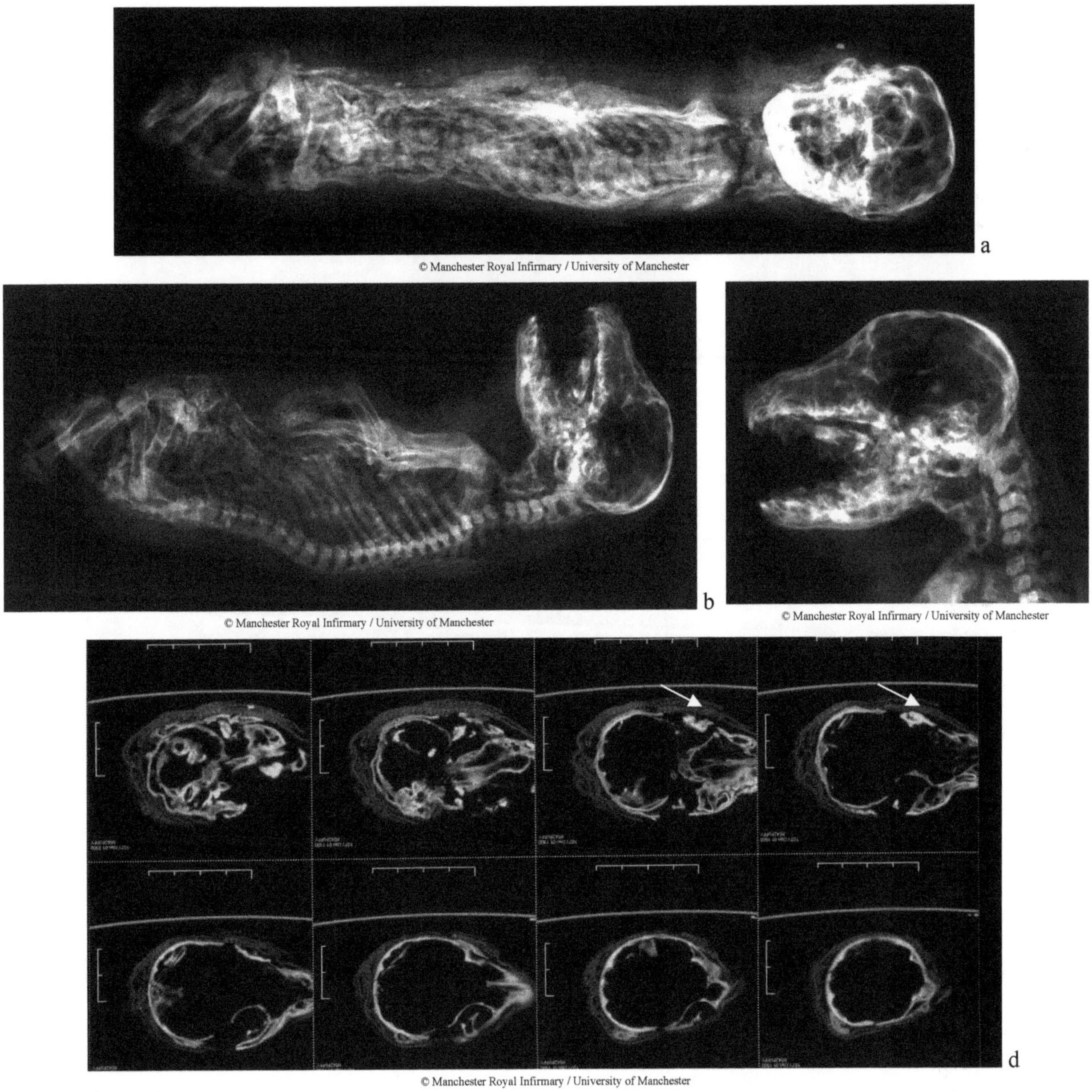

Fig. 9.3.1 – a) AP radiograph, b) Lat radiograph, c) region of interest highlighting the poor calcification of the dentition, and d) CT axial sections showing the radio-dense areas in the orbital sockets which may represent false eyes

question was raised as to whether they had been lost pre- or post-mummification. There is no visible opening through which the absent bones could have been lost. Desiccated brain matter is present at the rear of the cranial cavity. The CT images clearly show the growth plates on the lower limbs confirming that this is a juvenile animal.

9.3 – DOGS

MA3.1

Identification – Canine, small

Radiographic Report – XR – This mummy bundle contains the complete skeleton of a small canine demonstrating extensive loss of alignment in the spine. The bundle has been very tightly wrapped with the front paws down and the rear legs flexed and to the side. The mouth is wide open exposing the teeth which appear to be less dense than would normally be expected. The reason for this condition is unknown: it may be indicative of a diet low in calcium, or it is possible that unguents used during the mummification process could have caused the exposed teeth to take on this unusual appearance. No visible epiphyses can be confirmed with any certainty although there is possibly one on the distal femur (although this could have been caused by the superimposition of bones above). The animal is very small in proportion although it could be a mature animal. The animal had an elongated protruding snout.

CT – The CT study shows that some teeth may have been lost post-mortem and there is evidence to suggest the inclusion of false eyes shown as radio-dense areas within

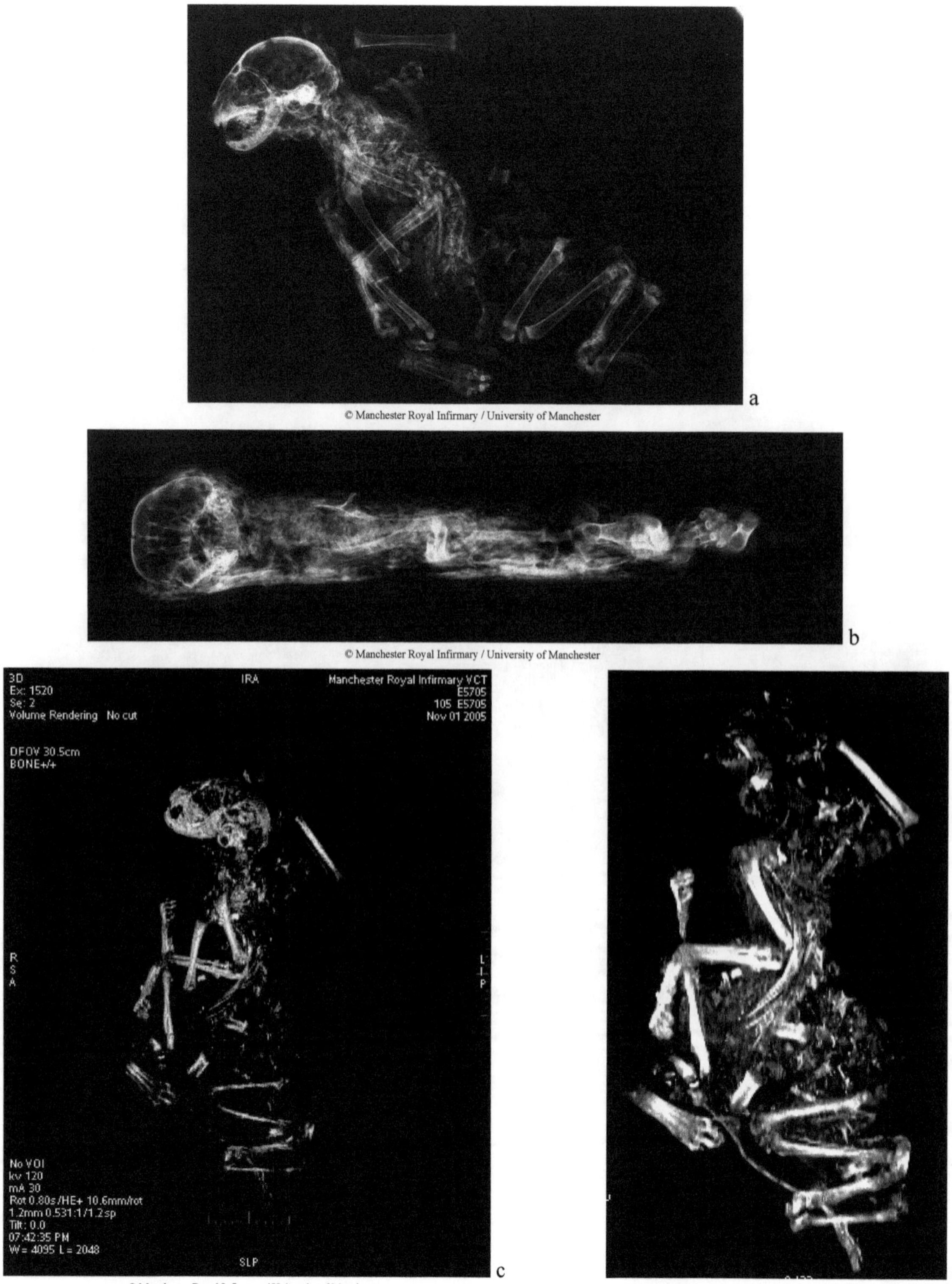

© Manchester Royal Infirmary / University of Manchester

© Manchester Royal Infirmary / University of Manchester

© Manchester Royal Infirmary / University of Manchester

© Manchester Royal Infirmary / University of Manchester

Fig. 9.3.2 – a) Lat radiograph, b) AP radiograph, c) volume rendering image created from CT data and d) coronal reformation. This specimen highlights the potential of computer software in the manipulation and visualisation of CT data which can greatly aid the interpretation process

the orbital sockets. The suspected false eyes are an interesting find which are not visible on the conventional radiographs. There are a number of radio-dense inclusions within the wrappings and there are visible patches close to the surface of the animal which could represent the presence of natron applied to the fur.

A post-mortem fracture is visible across the neck probably caused by the desiccated condition of the animal; however the external wrappings are not fractured suggesting that they must have been more pliable.

Conclusions – The CT study of this specimen added greatly to the information gathered during the preliminary radiographic investigation providing details on the bandaging and the possible inclusion of false eyes. The dimensions of the skull are more clearly visible on the CT images than they are on the conventional radiographs due to the angle at which the skull is lying within the mummy bundle.

MA3.2

'mummified cat'

Identification – Canine, juvenile

Radiographic report – XR- The images showed that this bundle contained the remains of a juvenile canine, with all the visible long bones appearing to be complete and to belong to one animal. There is evidence to suggest that the animal may have been allowed to partially decompose prior to wrapping suggested by the disarticulated nature of the skeletal remains.

CT – An anomaly to the vault of the skull is clearly visible which is thought to be the anterior fontanelle which had not fused at the time of death, again indicating a juvenile animal.

In total, mummy MA2.6 yielded 236 individual scanned images, making it essential to use computers to select the areas of interest. In this case, Volume Viewer was used; however as the software was developed for diagnostic radiology of human patients it is not always easily applied to mummified animals. It was necessary to experiment with thresholds to view different components of the body more clearly; however this is complicated by the similar attenuation of mummified muscle and soft-tissue with that of bone.

Volume Viewer was used to create a variety of different images such as scout projections which are the initial images created by the CT scanner and gives the radiographer the first insight into the contents of the mummy bundle. A virtual 'fly-through' of the mummy bundle was compiled using the axial section images and any areas of interest identified were subsequently viewed in more detail.

Three-dimensional reformations were created from the axial images. These reformations could be viewed in any plane and colours can be used to make interpretation of different features easier. These reformations were an excellent way of getting a three-dimensional view of the contents of the mummy bundle.

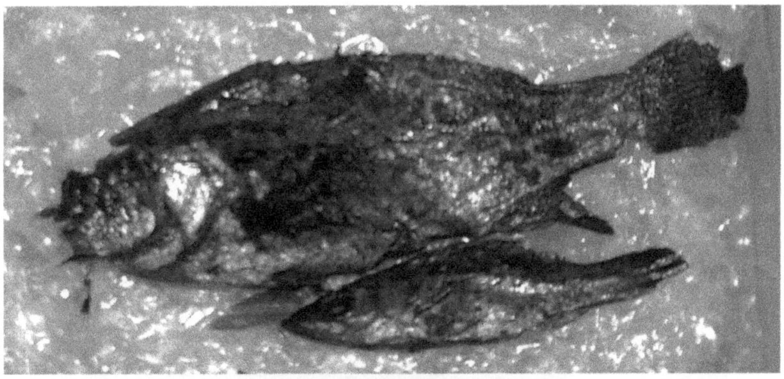

a

© University of Liverpool. Taken by Prof. J.E. Adams

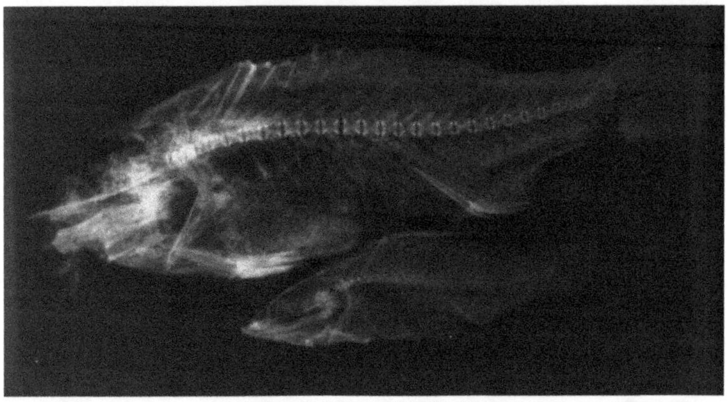

b

© Manchester Royal Infirmary / University of Manchester

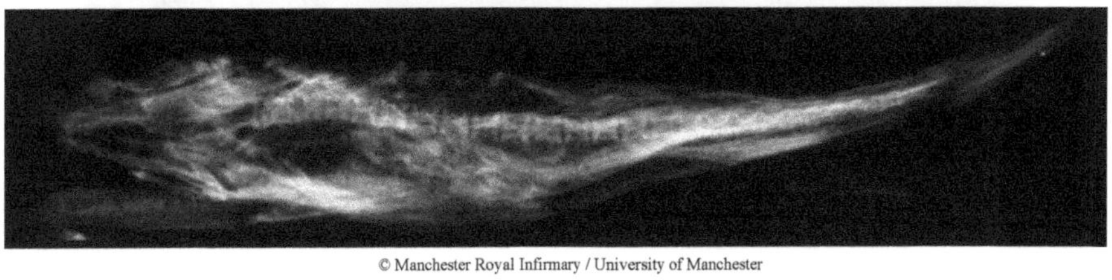

c

© Manchester Royal Infirmary / University of Manchester

Fig. 9.4.1 – a) Photograph, b) AP radiograph and c) Lat radiograph demonstrating the type of results achievable from poorly preserved mummies

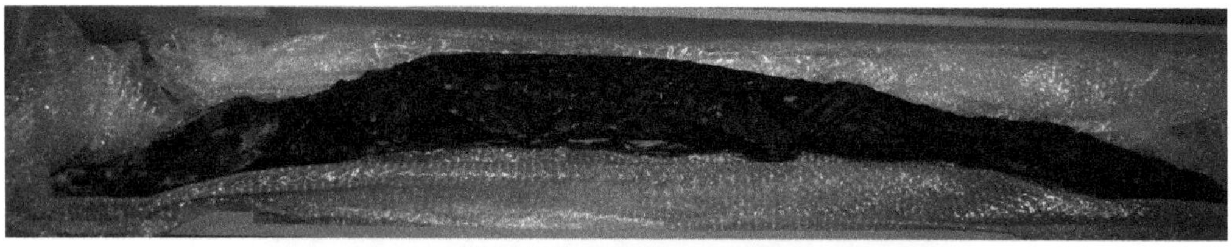
a

b

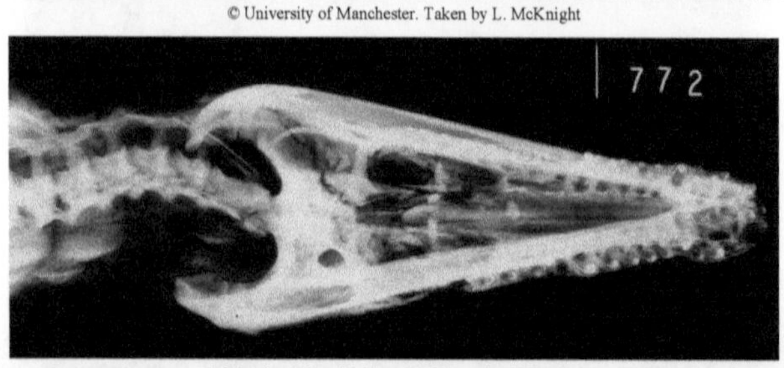

c

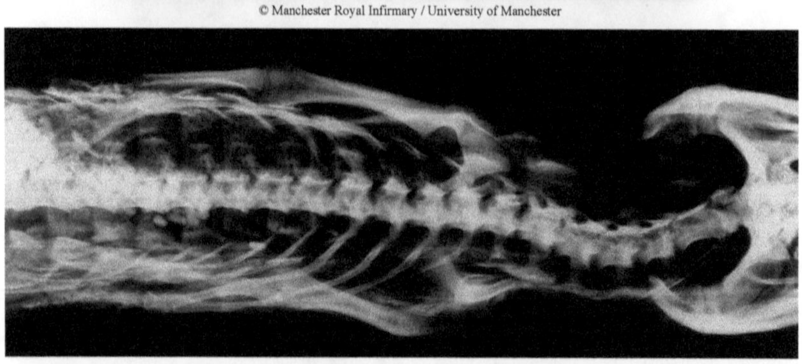
d

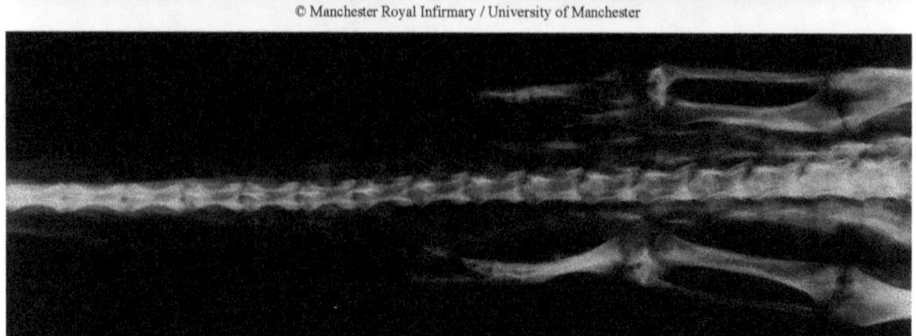
e

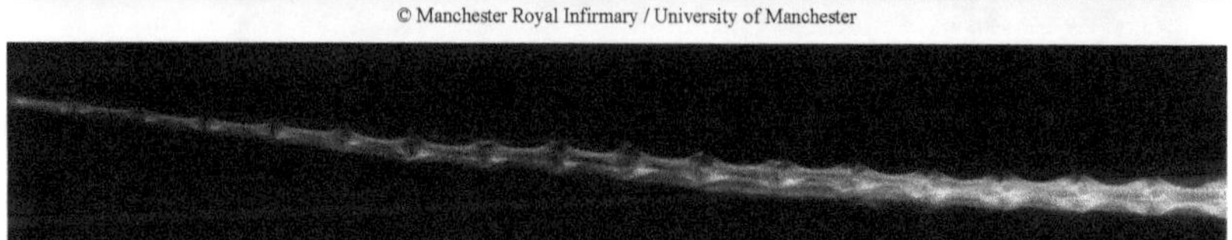
f

Fig. 9.5.1 – a) and b) Photographs, c) AP radiograph of the skull, d) cervical spine and upper limbs, e) lower limbs and f) tail. Photograph b) clearly shows the tar-like substance which can be seen covering the exterior of the mummy. Image c) shows the extensive skeletal damage to the peri-orbital bone. Image d) shows the presence of the ball-shaped mass in the stomach region

9.4 – FISH

MA4.1

'mummified fish'
(Presented in a box with MA7.5).

Identification – Fish

Radiographic Report – XR/CT – This specimen was presented for study in an incredibly poor state of preservation and it was necessary to conduct the analysis with the animal in the presentation box. No wrappings surrounding the mummy bundle are visible.

Conclusion – The radiographic analysis of this mummy yielded very little information. The specimen, along with numerous other mummified fish studied during this project, was simply in too poor a condition to gather useful information on the animal or the mummification technique employed. Another factor is the small dimensions of many of the fish mummies available for study which were too compressed to enable CT scanning to be effective. Mummified animals of this kind presented in such a poor condition are highly unlikely to produce any valuable information using CT scanning.

9.5 – CROCODILES

MA5.4

'Donated by M. E. Robinow between 1895 and 1896'

Identification – *Crocodylus niloticus* (Nile crocodile), mature

Radiographic Report – XR – This mummy had to be studied in five sections with overlap due to its large size (1697 mm in length). The mummy is complete and contains the skeleton of a crocodile with an estimated age at death of over five years. The spinal column appears to be intact with no fractures. There is a ball-shaped mass in the stomach with areas of opacity and evidence of granular fragments. There is little evidence to suggest that the crocodile was wrapped during mummification; however it is apparent that it was treated with a black tar-like substance which is visible on the skin of the mummy.

There is evidence of a linear transparency in the occipital region and disruption of the peri-orbital bone. There is evidence of fractures which has led to fragments of bone from the mandible or maxilla being displaced anterially. There has been extensive post-mortem tooth loss. The damage to the skull is thought to be post-mortem; however it is possible that this could represent the cause of death.

CT – The museum collection to which this particular animal belongs was not made available for study using CT scanning during the course of this project. The technique would be valuable in helping to determine how the injuries to the roof of the skull were sustained which could in turn suggest whether the injuries can be attributed to cause of death or whether they occurred post-mortem.

MA5.5

'Mummy thought to date to the Ptolemaic period. Donated to the museum by M. E. Robinow'

Identification – *Crocodylus niloticus* (Nile crocodile)

Radiographic Report – XR – The mummy bundle is wrapped in three sections of diagonal bandages with a straight line of fabric on top. The position of the eyes has been marked with protruberances made from linen. There is evidence of reeds within the bandages.

The radiographic images show that the mummy bundle contains the skulls of three crocodiles positioned in a line.

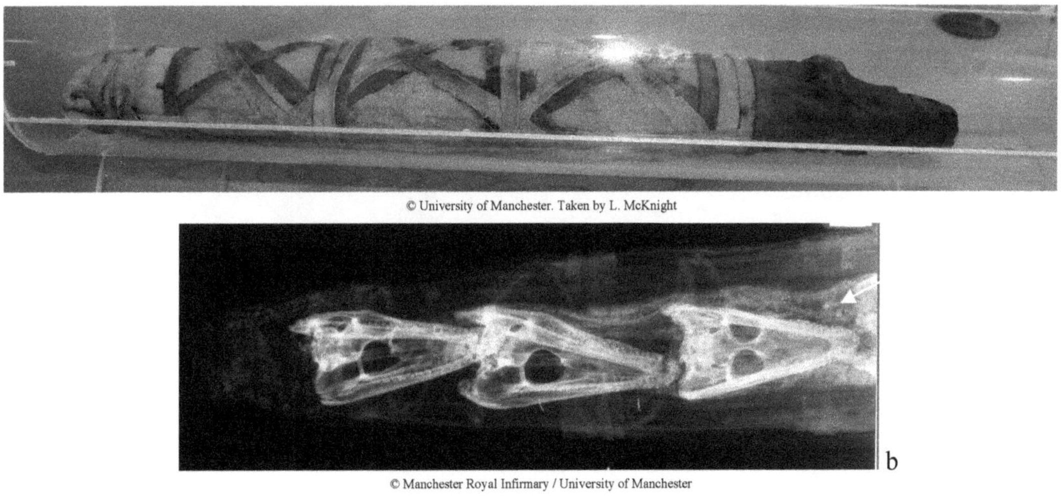

a

© University of Manchester. Taken by L. McKnight

b

© Manchester Royal Infirmary / University of Manchester

Fig. 9.5.2 – a) Photograph and b) AP radiograph showing the presence of three crocodile maxilla placed together within the mummy bundle. A further anomaly, thought to represent the remains of a juvenile crocodile or snake, can be seen lying on top of the two largest skulls. Reeds bound with thread are visible on the images

There is an area of interest to either side of the largest skull which may represent two smaller crocodiles, one lying on either side, or the remains of a snake.

CT – Again, this mummy belonged to the museum collection which was not available for study during this project; however it is believed that attempting the technique in the future would help to clarify the nature of the skeletal remains identified surrounding the largest skull.

9.6 – PSEUDO-MUMMIES

MA7.2

'Cone-shaped bundle thought to be a hawk or ibis. Egypt).

Identification – Pseudo-mummy

Radiographic Report – XR – This mummy bundle was found to contain no identifiable skeletal material. There is

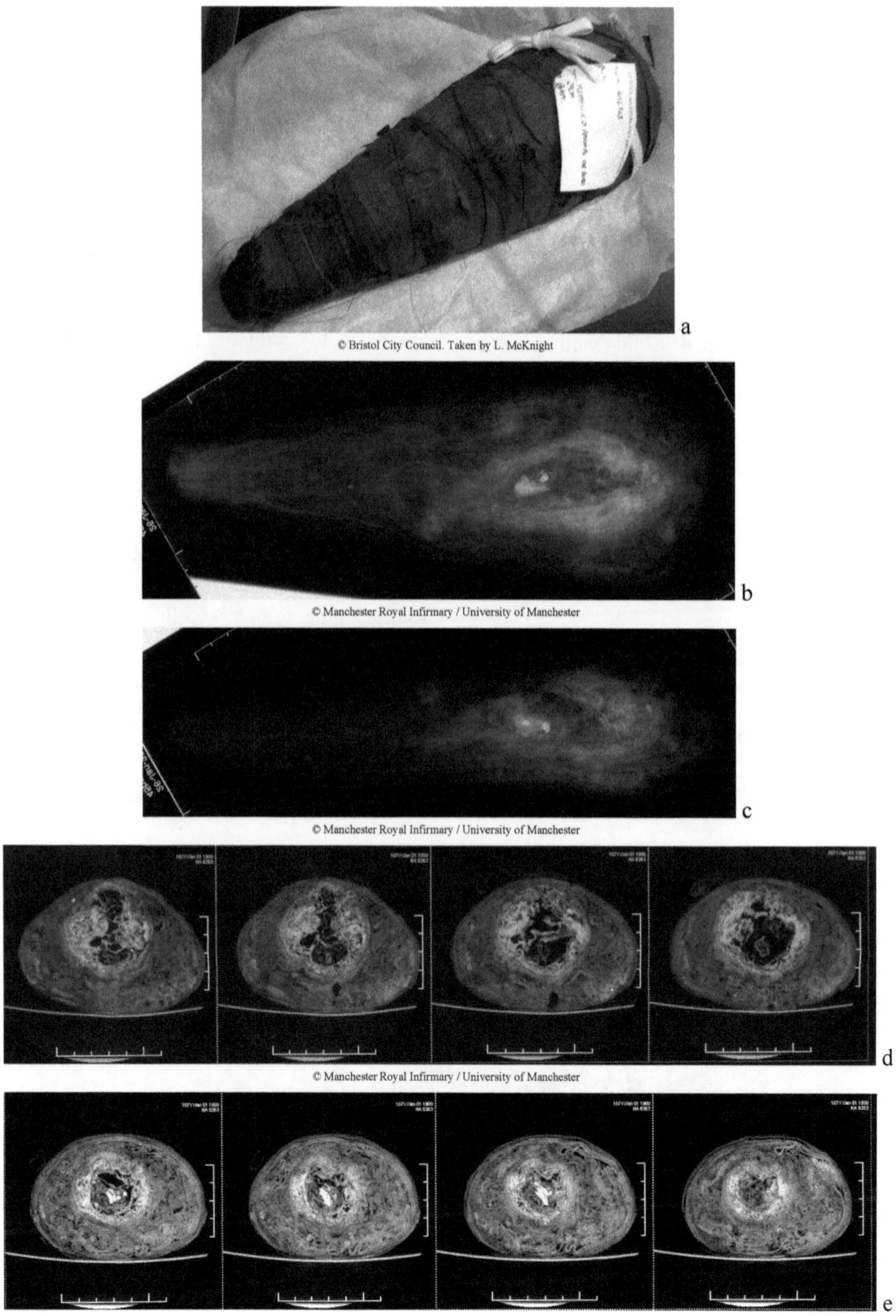

Fig. 9.6.1 – a) Photograph, b) AP radiograph, c) Lat radiograph, d) and e) CT axial sections showing no identifiable skeletal material. A number of radio-dense inclusions of unknown nature are visible

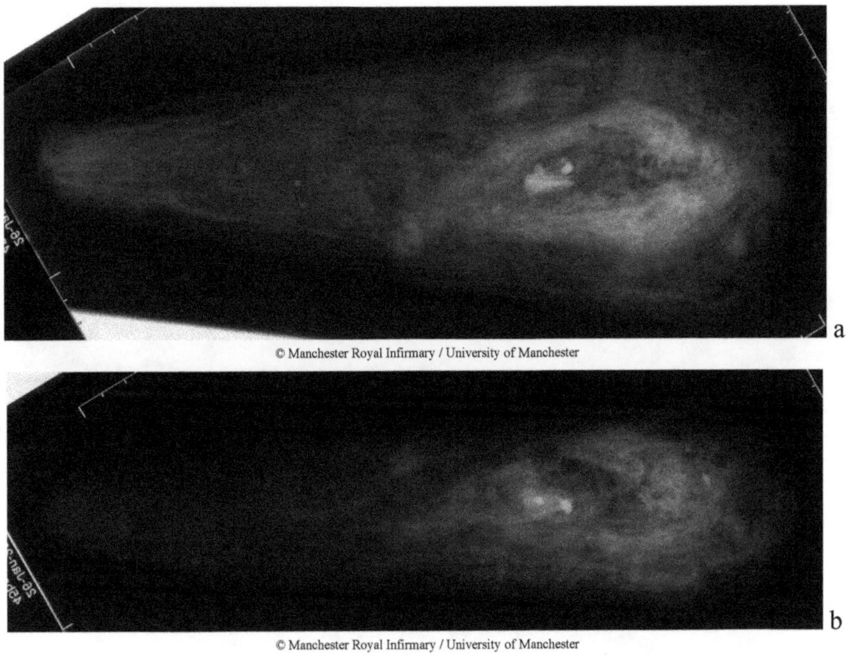

Fig. 9.6.2 – a) AP radiograph and b) Lat radiograph showing that the contents appear to have settled over time leaving air-filled voids

a bone-like anomaly which could represent a rib although this is not conclusive. There are a number of radio-dense inclusions which could be particles of metal, sand or stone.

The bundle displays a linear composition possibly created using feathers. It appears as though the contents were wrapped tightly, then re-wrapped again in another layer of bandages before being wrapped in a final layer.

There is a central void containing mud, stones and wax-like particles. A homogenous material with an appearance unlike wrappings seems to have been used to reinforce the body shape.

CT – The CT images show that there are no bones of any great size demonstrated through the lack of continuity of the fragments shown on the axial sections.

Conclusions – It can be difficult to ascertain the true size of bone fragments or other anomalies on the conventional radiographs simply because their margins are not always clearly visible. The use of CT scanning enables the researcher to view through the mummy bundle at designated places and in doing this it is possible to see how large the anomalies actually are.

MA7.18

'Mummified ibis'

Identification – Pseudo-mummy

Radiographic Report – XR – This is a complete mummy bundle containing bone and egg-shell fragments mixed with granular material.

CT – The CT images show that the contents of the bundle have settled to the base with an air-filled void above. The wrappings appear in a linear pattern. There is a small collection of vertebrae which could be from the vertebral column of a snake.

Conclusions – This mummy yielded greater information when viewed using CT scanning as it was possible to gain a better insight into the nature of the wrappings, the settling of the material within the bundle and the presence of the suspected snake remains.

MA7.19

Identification – Pseudo-bird

Radiographic Report – XR – This is the complete mummy of a bird modelled in a standing position containing no skeletal material and no visible anomalies. The outer layers of bandages are clearly visible on the head. One false leg has become detached from the body.

CT – The CT shows that the eyes are made from fabric with an outer level made of a high density material. As the specimen contains no visible anomalies, CT was able to yield little further information about its construction.

MA7.20

Identification – Pseudo-bird

Radiographic Report – XR – This is a complete mummy bundle containing no skeletal material. There are two visible anomalies within the bundle which are either semi-calcified eggs or bundles of fabric containing

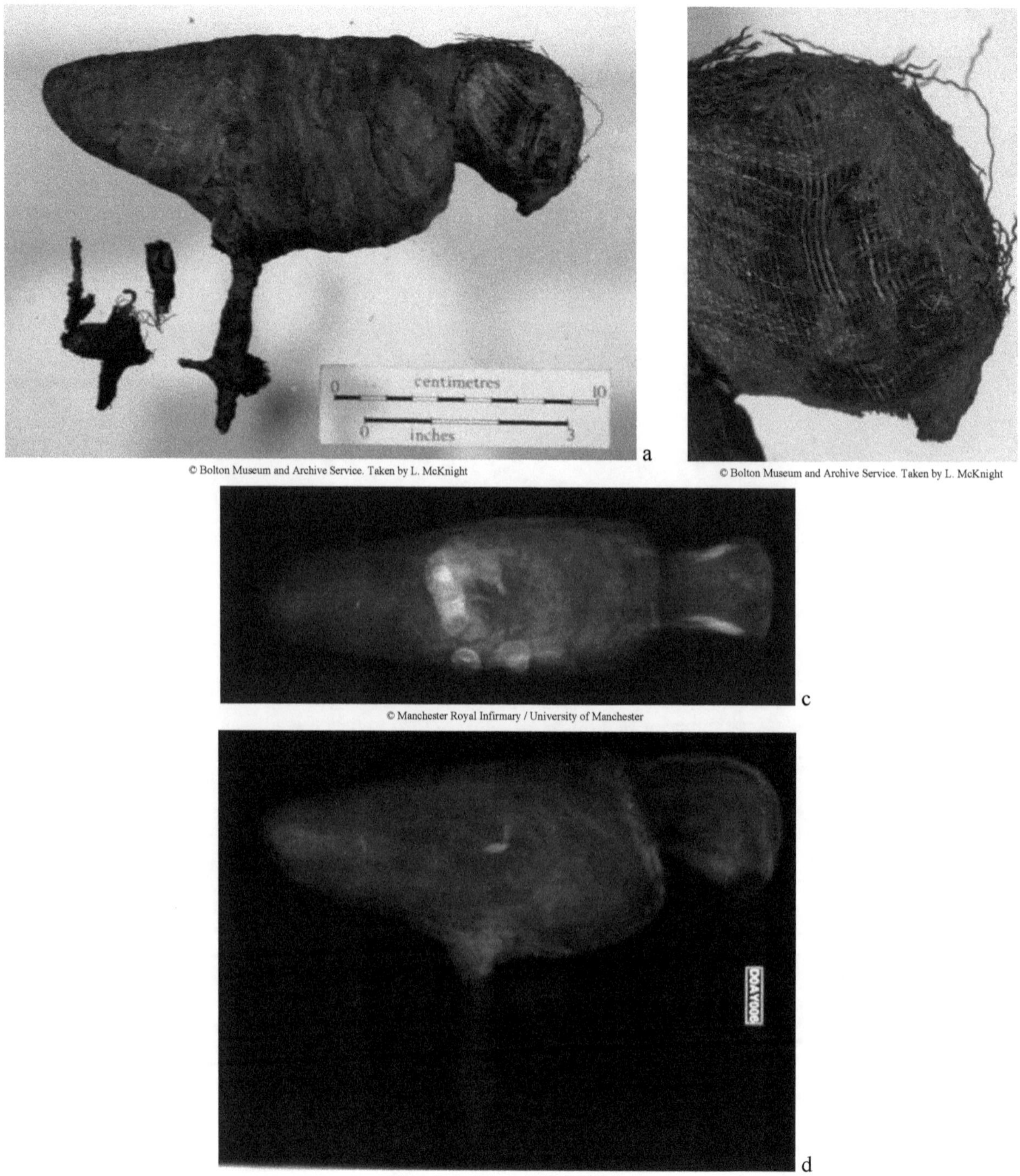

© Bolton Museum and Archive Service. Taken by L. McKnight

© Bolton Museum and Archive Service. Taken by L. McKnight

© Manchester Royal Infirmary / University of Manchester

© Manchester Royal Infirmary / University of Manchester

Fig. 9.6.3 – a) and b) Photographs, c) AP radiograph and d) Lat radiograph showing the complete lack of skeletal material within the mummy bundle

granular material. The mummy wrappings are 25mm thick at the deepest point.

CT – The CT images show that the bandages have been applied in three distinct layers which were not discernable on the conventional radiographs.

MA7.22

'Part of a group of eight crocodiles from Hawara'

Identification – Pseudo-crocodile

Radiographic Report – XR – This is a complete mummy bundle found to contain no skeletal material. There is a large tooth-shaped object visible on the images; its high density may indicate that it is made of stone or metal. Two metal pins are also visible, but these are thought to be modern additions used to fix the mummy to a presentation board. The body shape of the crocodile has been created from reeds laid longitudinally and bound with thin thread.

CT – The CT images clearly show the presence of the reeds within the bandages and the air trapped within them.

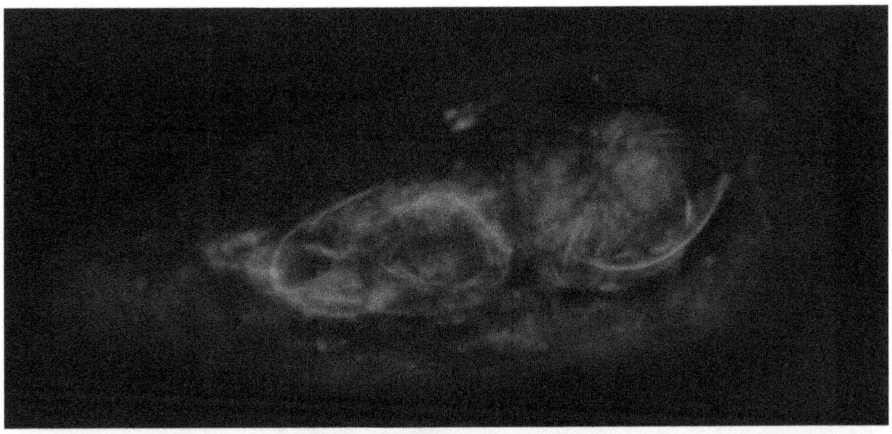

a

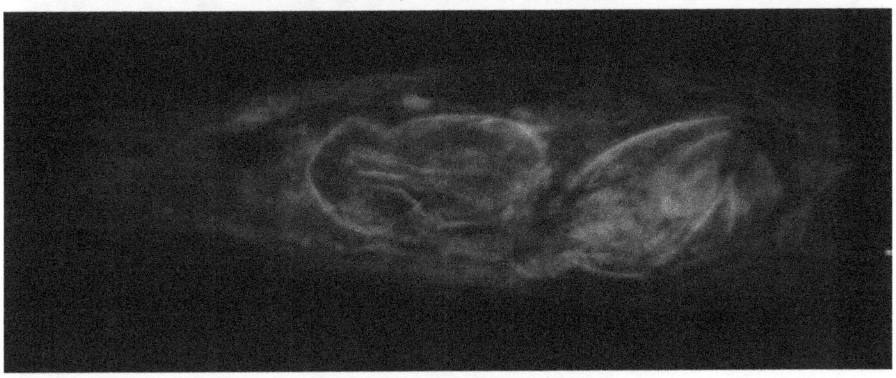

b

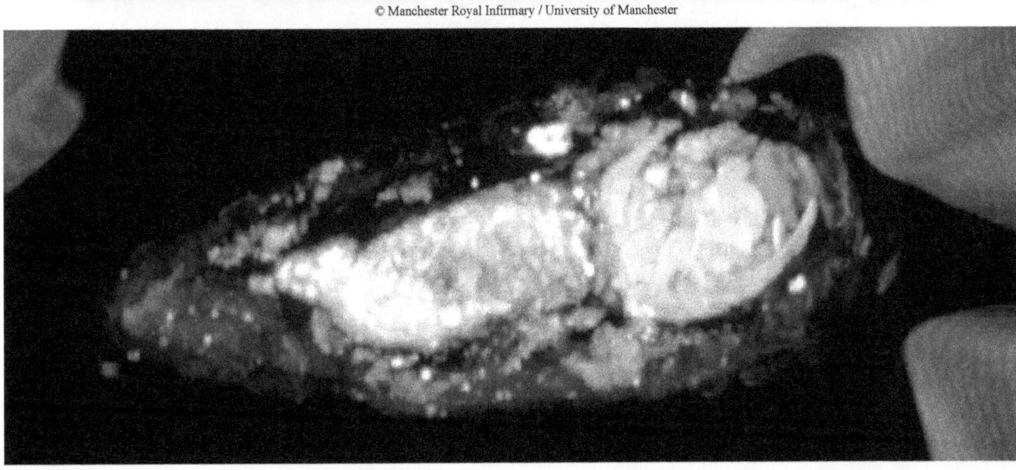

c

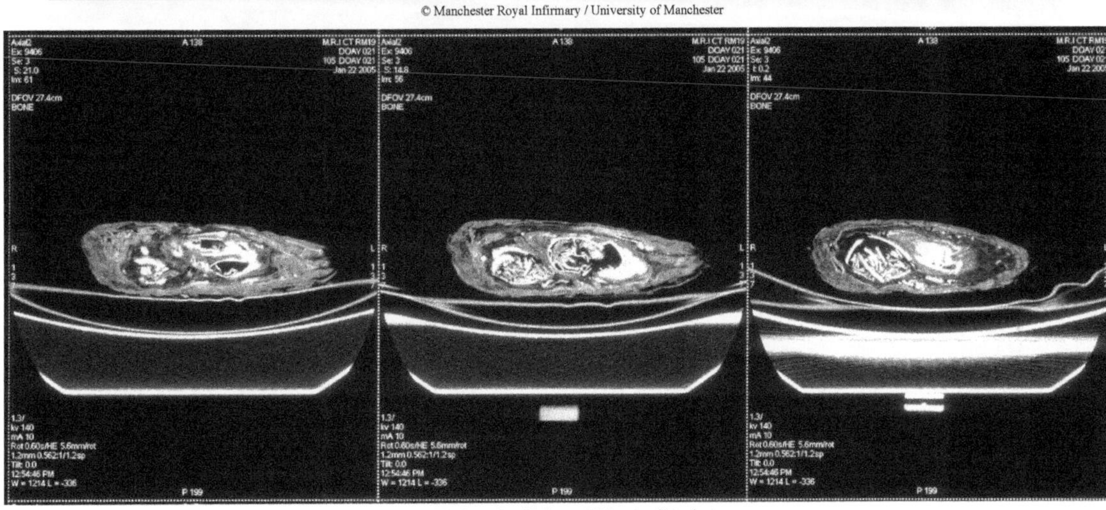

d

Fig. 9.6.4 – a) AP radiograph, b) Lat radiograph, c) CT reformation and d) CT axial sections showing that the bundle contains what are thought to be two semi-calcified eggs. The eggs have become compressed and fragmented over time due to their fragile composition

© Manchester Royal Infirmary / University of Manchester

Fig. 9.6.5 – AP radiograph showing the presence of a radio-dense tooth-shaped object

a

© Manchester Royal Infirmary / University of Manchester

b

© Manchester Royal Infirmary / University of Manchester

Fig. 9.6.6 – a) AP radiograph and b) Lat radiograph showing the presence of egg-shell fragments

MA7.25

'Part of a group of eight crocodiles from Hawara'

Identification – Pseudo-crocodile

Radiographic Report – XR – This is a complete mummy bundle which was found to contain no skeletal material. There is evidence of egg shell fragments and two metallic objects possibly made from metal film or gilding within the bundle. The body shape of the crocodile has been created from reeds laid longitudinally and bound with thin thread.

CT – The dense objects within the bundle have an extremely high attenuation (1253.6) which is similar to that shown by cortical bone or metal. A 'region of interest' was created using computer software to analyse the density of these areas.

Conclusion – Although the technique of CT scanning added little further information in this case, it was possible to use the scan data and computer software to determine the density of the high attenuation object within the bundle.

Chapter Ten

CONCLUSIONS: A CRITICAL APPRAISAL OF THE RADIOGRAPHIC STUDY OF MUMMIFIED ANIMALS

10.0 – INTRODUCTION

This chapter is intended to provide the reader with a critical appraisal of the radiographic techniques used during this research project with the aim of highlighting potential problems which are then not duplicated in future projects. The two radiographic techniques used, X-ray and CT, are discussed below separately in terms of the positive and negative points that each possess. A summary section is included at the start of each sub-chapter highlighting the main issues discussed.

10.1 – CRITICAL APPRAISAL

Conventional Radiography (X-Ray)

- Non-invasive and non-destructive
- Used to locate anomalies
- Superimposition of structures
- Magnification/distortion of structures
- Blurring
- Industrial film versus medical/veterinary film

There can be little doubt that radiography occupies a pivotal role in the study of mummified animal remains as it is the only research technique which is completely non-invasive and non-destructive. This makes it a popular method of gaining an initial insight into the contents of a mummy bundle before deciding on whether further invasive biomedical sampling and testing should be attempted. X-ray should therefore be the primary research tool when dealing with mummified material. This can help to identify the presence and specific location of anomalies which may merit further investigation.

As discussed in previous chapters of this study, X-ray is not without its problems and limitations. The three major issues concerning the use of this technique are the superimposition of features within the mummy bundle, the magnification of structures caused by their distance away from the film, and the degree of blurring which occurs. All of these issues can lead to problems with the analysis of the radiographic images and the interpretation of the results. There is no way to resolve these problems as they are essentially an inherent limitation of the radiographic technique, but it is important to understand that they do occur and to be prepared to allow for them when making interpretations based solely on conventional radiographs.

The use of industrial radiographic films could be to the advantage of archaeologists and Egyptologists when studying ancient specimens as it is more suited to the type of material being investigated (O'Connor and O'Connor 2005, 71). During my own research on the Bolton Museum collection during 1999, I was fortunate enough to be provided with some industrial film which yielded excellent results given that the radiographic equipment used was not of the highest quality and that the images were developed by hand (Owen 2000). As discussed earlier, the confines of this research project did not facilitate the use of industrial film as the imaging took place in a busy NHS hospital using their own films and automated developing units.

A number of studies have been conducted to attempt to establish the amount of distortion which is present when using X-ray (Pedersen 1997; Owen 2000). A study by Pedersen involved the direct measurement of fish bones compared to their measurement on radiographs. The conclusion was that the measurements taken from the radiographs were reliable and that it was quicker to measure them in this way rather than using the skeletal remains (Pedersen 1997, 277).

A research experiment carried out by the author in 1999 used radiographic images of the mummified remains as direct access to the bones for measurement was not feasible. In this study a mummified bird was radiographed and the bones measured from the image. The same bird was then radiographed again with a metal ruler placed on top of the mummy bundle at a height of 50mm above the radiographic film. The degree of magnification in the metal ruler amounted to thirteen percent. Bones and structures which were in the mummy bundle were closer to the film and therefore the degree of magnification would have been less (Owen 2000, 24).

A number of specimens studied during the current research project would have included structures at heights greater than the 50mm benchmark used in the experiment. In these cases it would seem reasonable to suggest that a fairly significant degree of magnification would be noted. Without dissecting the mummy bundles and measuring the bones directly, this statement is merely speculative.

The use of bone measurements taken from radiographs have been a common feature in previous research into mummified animals; however, no bone measurements were taken during this research project as it was thought to be unnecessary given the specific questions being asked. Studies which rely heavily on bone measurement to achieve accurate species identification, ageing and sexing, may encounter problems through these issues of magnification. In this project, comparative skeletal material was used to attempt to identify the animal species; however, the problems encountered were associated more with the skeletal collections themselves and the fact that the animals in question were generally modern European species groups and therefore possessed slightly different dimensions from the ancient specimens.

Computed Tomography (CT)

- Non-invasive and non-destructive
- Viewing/interpretation of DICOM images
- Eliminates magnification and superimposition
- Computer manipulation

CT is a more recent advance in radiographic technology and uses far more complex processes to gather a vast amount of information about the object being studied. A number of issues arise when using this technique as a research tool: firstly, the sheer volume of data produced from even the smallest animal mummy can be counter-productive and it is necessary to be extremely selective as to the images (if any) that are chosen for interpretation and illustration.

Secondly, the data produced through the scanning process are in DICOM images and it is necessary to have access to a computer program such as K-PACS which enables the images to be viewed off-site on a personal computer (PC), if the CT manufacturer's workstation is not available for interrogation and reformatting of the large number of transverse axial images acquired. Problems were encountered during this research project with the use of such programs on normal desk-top PCs. The computer equipment used in hospitals has the ability to run high volumes of images through these programs at speed to gather the desired information. A secondary point to consider is that the software is expensive to purchase and complicated and time-consuming to learn to use optimally.

CT offers the researcher a vastly increased ability to manipulate the resulting images over conventional radiographs using technologically advanced computer software. Problems generally arise through the complexity of the software and the level of knowledge required to operate it; however, as has been demonstrated in this project, interesting results can be achieved given time and patience.

The major issue concerning computer manipulation involves the use of thresholds to differentiate between components of the body on the images. Medical software allows these to be selected and isolated with different colours being introduced to aid visualisation and interpretation. In theory, this tool can be applied to the images of mummified remains; however in practice the reality is far more complicated. The process of desiccation and the use of embalming treatments have the effect of altering the structural composition, and so the attenuation of the tissues of the corpse. For example, mummified soft-tissue begins to resemble bone. This can make it difficult, or even impossible, to differentiate between anatomical features on manipulated 3D images.

10.2 – GENERAL SUMMARY

To conclude, I believe that this research project has helped to formulate a methodological approach to the study of mummified animal remains through the use of X-ray and CT. The 127 animal mummies studied revealed interesting information on the animal species and the mummification techniques employed by the ancient Egyptians to preserve the animals for eternity. The radiographic techniques used compliment each other in the majority of cases, with the exception of specimens which were found to be unsuitable for study using CT. These specimens tended to be those which had survived in a poor state of preservation or those of small dimensions, in which the spatial resolution of CT was a limitation.

10.3 – FUTURE WORK

This research project essentially represents only a small case-study of what could be achieved by the examination and scientific study of mummified animals. The lack of previous studies utilising scientific techniques results in this material being 'unchartered territory' for Egyptologists and scientists alike. I have made a few suggestions below for future studies which would be beneficial to increasing our understanding of the practice of animal worship and mummification in ancient Egypt.

This list is by no means exhaustive and represents ideas that have come to light during both my research and that of others.

Future research possibilities using radiographic techniques:-

1. This study investigated 127 animal mummies which, although not a small number, represents a tiny percentage of those in museum collections worldwide. There would be value in more animal mummies being studied using imaging techniques over the coming years to increase understanding and awareness of the material.

2. Animal mummies that have been studied in the past using radiography alone and those which revealed evidence of interesting anomalies in the process, should be studied again using CT.

3. Advances in radiographic technology are constantly being made and I believe that, as the first method of analysis in the study of this material, it is important to utilise such improvements where possible. As techniques improve, it may be possible for mummies such as those studied in this research project to be studied further with imaging to enhance knowledge and understanding.

4. It would be beneficial to the future study of mummified material for a dedicated computer program to be designed which has the capacity to deal with the DICOM images produced during the CT process. A tailor-made program would be far more suited to this type of material than the software programs in use in clinical practise in hospitals, and it could be expanded to include a cataloguing tool to organise images in specimen order.

5. It would be interesting to ascertain the differences between the use of industrial and medical radiographic film and developing equipment when studying ancient material. It may well be that the differences are minimal, but this would nonetheless be a worthwhile experiment.

Future research using alternative biomedical techniques:-

1. It would be interesting to attempt to apply histological methodologies which have already yielded successful results on mummified human material to animals. This technique could potentially identify pathological conditions present in the remaining mummified soft tissue which are unlikely to be evident with radiography. It is a method not without its technical problems as it is invasive and requires the removal of samples from the mummy bundle.

2. The retrieval of ancient DNA from mummified animals is an area of study which has not been widely attempted to date. In animals, the use of this technique could provide information on the genetic history of the breeds commonly mummified by the ancient Egyptians. This could then be used to trace demographic movements and migratory patterns which could demonstrate changes in environmental conditions.

3. Radiocarbon dating techniques can be used to suggest a date range within which an object belongs. The technique relies on the removal of samples which have not been contaminated in any way – something which is often not possible in the case of mummified animal specimens that have been subjected to years of poor storage and handling. The method has proved useful in confirming the date range for objects which were thought to belong to a certain historical period.

4. Mass spectrometry techniques could be utilised to identify the origin and chemical composition of substances used during the mummification process. This would prove useful in the case of the resins and oils applied during the treatment of animals by identifying the parallels with substances known to have been used on human mummies.

There can be little doubt that research into animal mummification is very much in its infancy and there are undoubtedly numerous areas for potential study. As scientific advances are made, any new methods could potentially be applied to this material. Essentially it would be encouraging to think that one day in the not-to-distant future Egypt's animal dead will be studied to the same rigorous scientific standards that have been applied to the human mummies.

10.4 – CONCLUSION

I hope that researchers seeking to increase our understanding of animal mummification will adopt the methodological approach used in this study for future projects. It is undoubtedly a methodology which is sure to evolve steadily over the coming years with the refinement of existing technology and the introduction of new techniques.

The decision to study ancient Egypt's animal dead, rather than the human remains, was one made some years ago whilst undertaking my undergraduate degree and it is a choice that I have never regretted. I believe that these animals have much to inform modern researchers about the lives they lived, the way they died and the processes they were subjected to after death. They constitute the material remains of a large and complex belief system and there can be little doubt that they hold many secrets which would serve to enlighten the modern world about our ancient ancestors. I look forward with anticipation to future projects in this area which I am sure will follow.

APPENDICES

1.0 – TABLE OF ALL MUMMIES STUDIED

MA Number	Museum Number(s)	Museum	Species
MA1.1	Ha6364	Bristol	Bird
MA1.2	135TET	Bristol	Bird
MA1.3	Ha6367	Bristol	Bird
MA1.4	H544	Bristol	Bird
MA1.5	Ha6359	Bristol	Bird
MA1.6	E5424	Liverpool	Bird
MA1.7	E5406	Liverpool	Bird
MA1.8	E5408	Liverpool	Bird
MA1.9	48/00	Liverpool	Bird
MA1.10	MM3, 6035	Manchester	Bird
MA1.11	MM17, 9371	Manchester	Bird
MA1.12	MM18, 9428, 9248	Manchester	Bird
MA1.13	MM19, 9429	Manchester	Bird
MA1.14	MM21, 11294, Robinow 70	Manchester	Bird
MA1.15	MM22, 11295, 22944, Robinow 69	Manchester	Bird
MA1.16	MM23, 1971.21	Manchester	Bird
MA1.17	MM27, 1981.576	Manchester	Bird
MA1.18	MM29, A957	Manchester	Bird
MA1.19	MM31, 4295	Manchester	Bird
MA1.20	MM40, 6098	Manchester	Bird
MA1.21	MM41, 11501	Manchester	Bird
MA1.22	Rosalie's Bird	Private	Bird
MA1.23	LX19a	Stkp MBC	Bird
MA1.24	DOAY002, 1:1983:137	Bolton	Bird
MA1.25	DOAY003, 45:1913:16	Bolton	Bird
MA1.26	DOAY004, 1:1983:155	Bolton	Bird
MA1.27	DOAY005, 1:1983:159	Bolton	Bird
MA1.28	DOAY007, 1:1983:156	Bolton	Bird
MA1.29	DOAY008, 1:1983:69a	Bolton	Bird
MA1.30	DOAY010a, 1:1983:69c	Bolton	Bird
MA1.31	DOAY011, 1:1983:285:1	Bolton	Bird
MA1.32	DOAY012, 1:1983:285:2	Bolton	Bird
MA1.33	DOAY014, A:126:1968	Bolton	Bird

MA Number	Museum Number(s)	Museum	Species
MA1.34	DOAY016, 1:1983:88	Bolton	Bird
MA1.35	DOAY036	Bolton	Bird
MA2.1	Kitten mummy	Bristol	Cat
MA2.2	Ha6357	Bristol	Cat
MA2.3	Ha6358	Bristol	Cat
MA2.4	Ha6356	Bristol	Cat
MA2.5	Ha6361	Bristol	Cat
MA2.6	Chester mummy	Liverpool	Cat
MA2.7	E5425	Liverpool	Cat
MA2.8	MM13, 6842	Manchester	Cat
MA2.9	MM14, 1977.1162	Manchester	Cat
MA2.10	MM15, 11123, Robinow 68	Manchester	Cat
MA2.11	MM16, 12015, 22948	Manchester	Cat
MA2.12	MM25, 12012, 22945, EG9	Manchester	Cat
MA2.13	MM28, A957	Manchester	Cat
MA2.14	MM34, 12016	Manchester	Cat
MA2.15	MM36	Manchester	Cat
MA2.16	LX18.1	Stkp MBC	Cat
MA2.17	LX18.2	Stkp MBC	Cat
MA2.18	DOAY034, 1:1983:87	Bolton	Cat
MA2.19	DOAY035, 91,890.00	Bolton	Cat
MA2.20	? No number	Bolton	Cat
MA2.21	skull	Bolton	Cat
MA2.22	E5705	Liverpool	Cat
MA3.1	Ha542	Bristol	Dog
MA3.2	MM12	Manchester	Dog
MA4.1	E5704a	Liverpool	Fish
MA4.2	E5706	Liverpool	Fish
MA4.3	E5707	Liverpool	Fish
MA4.4	E5708	Liverpool	Fish
MA4.5	E5709	Liverpool	Fish
MA4.6	E5710	Liverpool	Fish
MA4.7	E5711	Liverpool	Fish
MA4.8	E5712	Liverpool	Fish
MA4.9	debris	Liverpool	Fish
MA4.10	E5714	Liverpool	Fish
MA4.11	MM30	Manchester	Fish
MA4.12	DOAY009, 1:1983:69b	Bolton	Fish
MA4.13	DOAY020, 1:1983:?	Bolton	Fish
MA4.14	DOAY021, 10.1920.9	Bolton	Fish
MA5.1	MM1, 100/1, EA19/1	Manchester	Crocodile
MA5.2	MM2, 100/2, EA19/2	Manchester	Crocodile
MA5.3	MM7, 7892	Manchester	Crocodile
MA5.4	MM9, 1772	Manchester	Crocodile
MA5.5	MM10, 12008, 22941, EG95	Manchester	Crocodile
MA5.6	MM11, 12010, 22943	Manchester	Crocodile
MA5.7	MM32, Robinow 72	Manchester	Crocodile

MA Number	Museum Number(s)	Museum	Species
MA5.8	MM38, 134944	Manchester	Crocodile
MA5.9	DOAY025, 42.12	Bolton	Crocodile
MA5.10	DOAY031, 58.1911f	Bolton	Crocodile
MA5.11	DOAY010b	Bolton	Crocodile
MA5.12	DOAY022, 1:1983:95	Bolton	Crocodile
MA5.13	DOAY023a, 1:1983:77	Bolton	Crocodile
MA5.14	DOAY023b, 1:1983:77	Bolton	Crocodile
MA5.15	DOAY023c, 1:1983:77	Bolton	Crocodile
MA5.16	DOAY024	Bolton	Crocodile
MM6.1	MM8, 3005	Manchester	
MA6.2	MM42, 6033a/b	Manchester	Shrew
MA6.3	MM43, 1195	Manchester	Snake
MA6.4	STOPM.1992.432	Stkp MBC	Rat
MA6.5	DOAY018, A:128:1968	Bolton	
MA6.6	DOAY017, 1:1983:292	Bolton	
MA6.7	DOAY019, 1:1983:292	Bolton	
MA7.1	Ha6362	Bristol	Pseudo
MA7.2	Ha6363	Bristol	Pseudo
MA7.3	E5703	Liverpool	Pseudo
MA7.4	E5409	Liverpool	Pseudo
MA7.5	E5704b	Liverpool	Pseudo
MA7.6	MM4, 6293	Manchester	Pseudo
MA7.7	MM5, 5373c	Manchester	Pseudo
MA7.8	MM6, 5373d	Manchester	Pseudo
MA7.9	MM20, 11293, Robinow 71	Manchester	Pseudo
MA7.10	MM24, 6034	Manchester	Pseudo
MA7.11	MM26, 12013, 22946, EG8	Manchester	Pseudo
MA7.12	MM33, 134944	Manchester	Pseudo
MA7.13	MM35, 9849, EA16/2	Manchester	Pseudo
MA7.14	MM37, EA16/2	Manchester	Pseudo
MA7.15	MM44, 6032	Manchester	Pseudo
MA7.16	MM45, 11296, Robinow 67	Manchester	Pseudo
MA7.17	MM46, 5373b	Manchester	Pseudo
MA7.18	DOAY001, 1:1983:288	Bolton	Pseudo
MA7.19	DOAY006	Bolton	Pseudo
MA7.20	DOAY013	Bolton	Pseudo
MA7.21	DOAY015, 15.88	Bolton	Pseudo
MA7.22	DOAY026, 58.1911a	Bolton	Pseudo
MA7.23	DOAY027, 58.1911c	Bolton	Pseudo
MA7.24	DOAY028, 58.1911c	Bolton	Pseudo
MA7.25	DOAY029, 58.1911d	Bolton	Pseudo
MA7.26	DOAY030, 58.1911e	Bolton	Pseudo
MA7.27	DOAY033, 58.1911h	Bolton	Pseudo
MA7.28	DOAY032, 58.1911g	Bolton	Pseudo
MA8.1	Ha6370	Bristol	Oddity
MA8.2	? baby / doll	Bolton	Oddity
MA8.3	? Conical doll	Bolton	Oddity

2.0 – MANCHESTER ANIMAL MUMMY PROJECT CATALOGUE

This catalogue is intended to provide supplementary information which was not included in the main text of this research project. It combines information on the external aspects of the mummy bundles as well as a summary of the radiographic results. The results have been subdivided into X-ray and CT to enable the reader to assess the value of each method in individual cases. The catalogue has been arranged in individual collections, rather than in numerical order. Information given in italics relates to documentation/labelling found with or on the object at the time of study.

BRISTOL MUSEUM
Fourteen Specimens
Contact Details:
Bristol Museum and Art Gallery, Queen's Road, Bristol BS8 1RL.
Tel. 0117 922 3571 Fax. 0117 922 2047
Email. general.museum@bristol.gov.uk

MA1.1 – (Ha6364)
'finely wrapped in square design with false foot. Late Period? Egypt'
Identification – cf. Falcon. Nostrils similar to *Accipiter nisus* (Sparrowhawk).
Radiographic Report – XR – Bundle is fractured (separated) in the region of the tail. The skeleton is much smaller than the mummy bundle (the skull is placed within the modelled body, not in the head area – false head). The false head appears to have a radio-dense coating which could be resin or gilding. There appears to be an area of soft-tissue or feathers which is extraneous to the bird, but is included within the bundle wrappings (may be the tail area of another bird).

CT – Clearly shows that there is a thin layer (possibly even a single layer) of bandages around the mummy which appear radio-dense (coating of tar, paint or natron?). This bundle is then wrapped again to form a more uniform cylindrical shape with what appears to be rolls of bandaging on each side of the neck (for support?). This bundle then appears to have been rewrapped with the inclusion of denser material (possibly wax) laid intentionally in a row within the wrappings. The mummy has been given false head and feet. CT has been very informative on this mummy giving a better insight into the wrappings in particular.

MA1.2 – (135TET)
'Possible fake with rough bandaging and feet. Egypt'
Identification – Avian (decapitated skeleton – unidentifiable), insectivore (skull only)
Radiographic Report – XR/CT – The bundle appears to be mostly wrappings. Decapitated bird skeleton at the 'head end'. None of the larger bones are diagnostically avian. There are a mass of smaller bones and what appears to be the skull of an insectivore.

MA1.3 – (Ha6367)
'Wrapped in wide bandages with narrower ones on top. Thebes'
Identification – Avian? (Possibly pseudo-bird but included with birds because of presence of bone and feather fragments)
Radiographic Report – XR – The X-rays show bone-like anomalies and feathers. This looks like a bundle of feathers that have been modelled to form the shape of the mummy and a modelled head with layers of dense bandaging with a beak and ears. There is a particle of a radio-dense substance mid-way.
CT – Produced a vast amount of information on the way the mummy has been formed. There appear to be unidentifiable bone fragments and there is evidence of a tar-like coating.

MA1.4 – (H544)
'Rough bandaging. Mrs P. A. Fry 1902'
Identification – cf. Falcon
Radiographic Report – XR/CT – Externally the bundle is long, but the area containing the skeleton appears to be relatively plump. The tail is long (indicative of cf falcon) and the beak is protruding from the bandages. The legs are flexed up and the claws are not visible (unusual position for a bird). The neck is displaced and there is a residue in the chest cavity. Consensus is that it is the skeleton of a falcon, but that it is positioned strangely within the mummy bundle leading to a slightly elongated appearance on the films.

MA1.5 – (Ha6359)
'Roughly bandaged. Egypt'
Identification – cf. Falcon. Possibly *Falco tinnunculus* (Kestrel)
Radiographic Report – XR/CT – The tail feathers are folded back on themselves and the dense appearance suggests that they may have been coated in resin. The mummy does not seem to have been wrapped as tightly as the other mummies. There are thoracic contents present within the body cavity.

MA2.1 – (Kitten Mummy)
'bandaged with facial features painted on. Egypt'
Identification – feline, juvenile
Radiographic Report – XR – Largely disarticulated skeleton. The skull vault has collapsed inwards leaving a mass of bone fragments in the back of the cranium. The neck has been broken (post-mortem) possibly due to the plastic strap keeping the kitten strapped to the board. There appears to be a small radio-dense inclusion within the bundle (metallic flake, sand or stone).

CT – CT clearly shows the cloth ears and eyes. There is evidence that the bundle has been wrapped twice, but no evidence for radio-dense coating. 3D CT of the skull may help to clarify what has occurred.

MA2.2 – (Ha6357)
'Carefully wrapped in different coloured bandages, arranged in a striped pattern. Facial features painted on. Beni Hasan. Mrs A. E. Hudd'

Identification – Feline, juvenile
Radiographic Report – XR – The skeleton fills the mummy bundle completely and is intact and fully articulated. There is a dense area on the ear that protrudes from the bandages which may be resin used during the mummification process. There are a number of epiphyses visible and the secondary dentition is present high in the tooth sockets suggesting that the animal was immature. The hind legs are flexed upwards in the 'normal' manner.

CT – CT shows that the orbital sockets are completely empty with no evidence for false eyes or dried orbital matter. There are three distinct layers of bandages. CT also clearly shows the wings of the scapuli and a nicely articulated skeleton.

MA2.3 – (Ha6358)
'Roughly bandaged, though the face is carefully modelled from bandages with 'buttons' of linen for eyes. Egypt'
Identification – Feline
Radiographic Report – XR – The skeleton shows extensive disruption of the skull with the displacement of the petrous bones towards the rear of the skull and disruption of the vertebral column. The thorax has collapsed inwards and some ribs appear in the neck region. The pelvis is not visible although there are some unidentifiable bony fragments. One forelimb is fractured, the other is intact. This mummy could represent a partially disarticulated skeleton that has been remodelled to appear more like a cat.

CT – The CT images shows evidence of rolls of fabric within the mummy bundle.

MA2.4 – (Ha6356)
'Bandaged with painted facial features. Beni Hasan. Mrs AE Hudd'
Identification –Feline, juvenile
Radiographic Report – XR – The petrus bones are in their correct anatomical position. The neck has been broken (post-mortem) – there is no evidence of twisting or displacement of adjacent bones which would be visible if the cat had been strangled. There is also a post-mortem fracture of the humerus. The tail is present and has been folded up along the spine. The facial bones are missing and can be seen at the rear of the cranial cavity, yet the wrappings appear to be intact over the face. Some epiphyses are visible suggesting that it is an immature animal.

CT – Images show bandaging clearly.

MA2.5 – (Ha6361)
'Roughly bandaged. Egypt'
Identification – Feline, juvenile
Radiographic Report – XR – Modelled to look like a complete cat, but only contains two front limbs. No skull visible. There is a post-mortem fracture of the humerus. The wrappings appear to be loose at the 'head-end' and there is a fracture of the mummy bundle in the neck region.

CT – The CT shows that this cat has been wrapped once, dipped in a substance then wrapped again. It appears as if the body of the cat could have been formed from an unknown material, with the bones of the front limbs being placed alongside in anatomical position.

MA3.1 – (Ha542)
'Rough bandaging, open mouth. Rev. Morris 1823. Egypt'
Identification – Canine, small
Radiographic Report – XR – Complete skeleton showing loss of alignment in the spine. Very tightly wrapped with the front paws down and the rear legs flexed and to the side. The mouth is wide open although the teeth that are present do not appear to be as dense as you would expect (reason unknown). CT shows that some teeth may have been lost post-mortem and suggests the inclusion of false eyes. There are no visible epiphyses (possibly one on distal femur although this could be superimposition) although the animal is very small (mature animal). The snout is very long and elongated.

CT – There are inclusions within the wrappings which appear radio-dense and there are patches close to the surface of the animal (natron in fur?). The false eyes are an interesting find which are not visible on the X-rays. A fracture across the neck (post-mortem) could be because the animal's desiccated condition caused it to snap, but that the external wrappings were more flexible and did not fracture. CT has added a lot of information about the bandaging. The dimensions of the skull are more clearly visible on the CT scans than they are on the X-rays because of the angle of the skull in the bundle.

MA7.1 – (Ha6362)
'Cone-shaped bundle thought to be a hawk or ibis. Egypt'
Identification – pseudo-mummy – 'floor sweepings'
Radiographic Report – XR/CT – Contains isolated unidentifiable bone fragments (nothing distinct for ID). There is an area thought to be resin soaked bandages. An air space is visible above the contents indicating that they have settled over time leaving a void.

The central void is likely to contain mobile contents – the floor sweepings probably move about within the cavity and settle at the lowest point. There are some anomalies which could be sticks or reeds positioned longitudinally to provide support and rigidity. The whole bundle seems to have been wrapped fairly loosely.

MA7.2 – (Ha6363)
'Cone-shaped bundle thought to be a hawk or ibis. Egypt)
Identification – pseudo-mummy
Radiographic Report – XR – No skeletal material clearly visible with the exception of what could be a rib. There are a number of dense inclusions (metal, sand or stone). The bundle has a linear composition – possibly feathers? It appears as though the contents were wrapped tightly, then re-wrapped again in another layer of bandages before being wrapped in a final layer. There is a central void containing mud, stones and wax-like particles. The

bundle seems to have been padded out with a homogenous material that does not look like the bandages.

CT – It would appear that there are no bones of any great size as there is no continuity of the fragments shown on the axial sections.

MA8.1 – (Ha6370)
'Indeterminate bird or animal mummy, in a coffin-shaped box thought to be 19th Century. Mummiform with rough human face in plaster or clay. Possibly a tourist fake? Egypt'
Identification – pseudo-baby
Radiographic Report – XR – Human faced mummy presented within a wooden coffin with a glass lid. The face appears to have been modelled from a substance similar to plaster of Paris. There has been some damage to the mid-section of the mummy. The fabric from which the body shape has been modelled is relatively dense which suggests that the bandages may have been soaked in resin prior to wrapping.

CT – CT shows damage to the mummy bundle well. Likely to be an ancient mummy deduced from the way the bandages have been applied in multiple layers in a similar fashion to others within the collection. The plaster face and coffin have probably been added in the Victorian period. There are no bones within the bundle whatsoever, but the bandages appear to have been dipped in a substance which appears radio-dense on the images. The threshold on the CT was set to high to eliminate the bandages at the foot end. This reveals the plaster 'face' and the central core of the mummy bundle.

SCHOOL OF ARCHAEOLOGY, CLASSICS AND EGYPTOLOGY, UNIVERSITY OF LIVERPOOL
Twenty Specimens
Contact Details:
12-14 Abercromby Square, University of Liverpool, Liverpool L69 7WZ
Tel. 0151 794 2467
Fax. 0151 794 2442
Email. saceadm@liv.ac.uk

MA1.6 – (E5424)
'Thought to be a mummified bird. Beni Hasan? c.2500 BC. In poor condition with the head separated from the body. The bandages are torn at the bottom end of the mummy and bones are showing through. It was found on Thurs. 12th Nov. 1992 by workmen clearing out the lofts of no. 11 Abercromby Square together with other mummified pieces, basket ware and pottery. No. 11 Abercromby Square had been part of the old Inst. of Archaeology before the 1940s, and the material had been put under the eaves in the roof space and forgotten about.'
Identification – Avian. Parts of a juvenile ibis and a smaller bird cf. Falcon
Radiographic Report – XR/CT – appears to contain odd bones of a large bird, but has no skeletal integrity. There are the distal phalanges (ends of the wings), pair of intact ulna, base of both distal humerii, fragments of both radii (wings). The tail section contains a fragment of the radius thought to belong to the same bird (size consistent with an ibis), with the distal tarsometatarsus (claw) of a smaller bird. There are Harris lines in the wing bones. Skull and vertebrae not present. The base of the back toe is isolated, but there are some disarticulated toes elsewhere.

MA1.7 – (E5406)
'Mummifed head, neck and body of a bird'
Identification – cf. Falcon. Possibly (*Falco tinnunculus*) Kestrel, with parts of a larger falcon, possibly lanner or saker
Radiographic Report – XR/CT – Disarticulated, jumble of bones. The wing, ulna, radius and humerus are articulated and the feathers would have given the shape. The pelvis and femur are articulated. Femur lost mid shaft. The other femur head is present, but is not articulated. Leg, corocoid and scapula may have been positioned in almost the correct position, but are not articulated. A section of the spinal column has been separated. It may or may not be from the same bird.

The skull and leg on the other image are unlikely to belong to the same bird as the proportions are different (bigger). Basal view of the skull shows extensive damage to the inside of the cranium. The mandible is present. Size is consistent with a larger falcon ie. peregrine or saker.

MA1.8 – (E5408)
'Mummified hawk in good condition'
Identification – cf. Falcon
Radiographic Report – XR – Complete bird mummy. Legs folded up over the chest towards the head – unusual position. Tail and wing feathers creating length of the bird, but no bones present in lower half of the bundle. The neck has been forced down on itself creating displacement and fore-shortening. The talons are complete. No visible body contents.

CT – Highlights unusual texture visible on the tail section of the mummy.

MA1.9 – (48/00)
'Parts of a mummy, probably a bird'
Identification – Avian
Radiographic Report – XR/CT – Contains one leg and wrapping only.

MA2.6 – (Chester Mummy)
'Small anthropoid mummy case containing a large cat with some additional bones. Late period. X-rayed by Professor Harrison of the Dept. of Anatomy in March 1973. Given to Grosvenor Museum, Chester, in 1888 by Rear Admiral Sir Richard Massie Blomfield. On long term loan to Liverpool since February 1974.'
Identification – Feline
Radiographic Report – XR – The mummy bundle was studied in sections due to its size with an overlap. Complete mummy bundle containing one cat (identified

dentally). Outer beaded layer which appears as a mesh on the images. Shows magnification of the bones farthest away from the film. There appears to have been post-mortem dismemberment. The lower portion of the mummy bundle contains the remains of the lower limbs, partially overlapping. The tail has been severed and can be seen lying towards the base of the mummy bundle. The images demonstrate the issue of magnification of the bones caused by X-ray.

MA2.7 – (E5425)
'Mummified cat with pleated bandages. Beni Hasan? c.2500BC. Poor condition, with the bottom of the bundle damaged and open. It was found on Thurs. 12th Nov. 1992 by workmen clearing out the lofts of no. 11 Abercromby Square together with other mummified pieces, basket ware and pottery. No. 11 Abercromby Square had been part of the old Inst. of Archaeology before the 1940s, and the material had been put under the eaves in the roof space and forgotten about.'
Identification – Feline
Radiographic Report – XR/CT – Beautifully modelled cat bundle containing a fragment of femur and some vertebrae. Circular area at lower end could be wax disk. Modelled face and ears.

MA2.22 – (E5705)
'Mummified cat'
Identification – Feline, juvenile
Radiographic Report – XR– Contains the decomposed remains of a kitten. Epiphyses are visible. All visible bones appear to be complete and to belong to one animal.

CT – An anomaly to the vault of the skull is clearly visible which is thought to be anterior fontanelle which had not fused at the time of death, again indicating a juvenile animal.

In total, mummy MA2.6 yielded 236 individual scanned images making it essential to use computers to select the areas of interest. In this case, Volume Viewer was used, however as the software was developed for diagnostic radiology of human patients it is not always easily applied to mummified animals. It was necessary to experiment with thresholds to view different components of the body more clearly; however this is complicated by the similar attenuation of mummified muscle and soft-tissue with that of bone.

Volume Viewer was used to create a variety of different images such as scout projections which are the initial images created by the CT scanner and gives the radiographer the first insight into the contents of the mummy bundle. A virtual 'fly-through' of the mummy bundle was compiled using the axial section images and any areas of interest identified were subsequently viewed in more detail.

Three-dimensional reformations were created from the axial images. These reformations could be viewed in any plane and colours can be used to make interpretation of different features easier. These reformations were an excellent way of getting a three-dimensional view of the contents of the mummy bundle.

MA4.1 – (E5704a)
'mummified fish'. (Presented in a box with MA7.5).
Identification – Fish
Radiographic Report – XR/CT – Very brittle mummy yielding little information on analysis. No wrapping visible.

MA4.2 – (E5706)
'Mummified fish in poor condition'
Identification – Fish
Radiographic Report – XR/CT – Skeleton of a fish with no skull present. Evidence for disarticulation of the skeleton and sand packing of the body cavity.

MA4.3 – (E5707)
'Mummified fish'
Identification – Fish
Radiographic Report – XR/CT – Decapitated fish skeleton with skull and distal end of tail lost. Not wrapped.

MA4.4 – (E5708)
'Mummified fish'
Identification – Fish
Radiographic Report – XR/CT – Contains the partial disarticulated skeleton of a fish. Not wrapped.

MA4.5 – (E5709)
'Mummified fish'
Identification – Fish
Radiographic Report – XR/CT – Contains the skeleton of a fish with no tail present. There is evidence for small bones and vertebrae in the stomach area.

MA4.6 – (E5710)
'Mummified fish'
Identification – Fish
Radiographic Report – XR/CT – Contains the disarticulated partial skeleton of a fish. Not wrapped.

MA4.7 – (E5711)
'Mummified fish'
Identification – Fish
Radiographic Report – XR/CT – Contains the disarticulated partial skeleton of a fish with the skull detached. Not wrapped.

MA4.8 – (E5712)
'Mummified fish'
Identification – Fish
Radiographic Report – XR/CT – Contains the skeleton of a fish. Thought to be complete, but extensive disarticulation. No visible bandages.

MA4.9 – (debris)
Identification – Fish
Radiographic Report – XR/CT – Contains the mangled remains of a fish. Not wrapped and extensively disarticulated.

MA4.10 – (E5714)
'Mummified fish'
Identification – Fish
Radiographic Report – XR/CT – Radiography shows a complete fish skeleton with a fracture through the bundle in the mid-section. There is a degree of skeletal displacement near the tail.

MA7.3 – (E5703)
Identification – Pseudo-mummy
Radiographic Report – XR/CT – No skeletal material present. Modelled from fabric.

MA7.4 – (E5409)
'Mummified eel?'
Identification – Pseudo-mummy
Radiographic Report – XR/CT – Catalogued as an eel from the outward appearance but contains no skeletal material. It appears to have been constructed from rolled fabric displaying concentric circles on the images.

MA7.5 – (E5704b)
Identification – Pseudo-mummy
Radiographic Report – XR/CT – Completely empty package containing no skeletal material.

BOLTON MUSEUM AND ART GALLERY
Forty-three mummies
Contact Details:
Bolton Museum and Archive Service, Le Mans Crescent, Bolton, Lancashire BL1 1SE
Tel. 01204 332211
Email. Museum.customerservices@bolton.gov.uk

MA1.24 – (DOAY002, 1.1983.187)
'Beni Hasan'
Identification – *Hieraaetus fasciatus* (Bonelli's Eagle) or *Hieraaetus pennatus* (Booted Eagle).
Radiographic Report – XR/CT – Articulated femur and tibiotarsus displaying healthy bone cortex. The specimen is not wrapped and is thought to belong to a mature bird.

MA1.25 – (DOAY003, 45.1913.16)
'Femora of Sacred Ibis from Abydos'
Identification – *Threskiornis aethiopicus* (Sacred Ibis)
Radiographic Report – XR/CT – Two complete ibis femora with one displaying a well-healed fracture to the shaft at an angle of ninety degrees. The specimen is presented in a box along with what appears to be egg-shell fragments. The femora are not wrapped.

MA1.26 – (DOAY004, 1.1983.155)
Identification – *Falco tinnunculus* (Kestrel)
Radiographic Report – Incomplete mummy of a mature kestrel displaying a post-mortem fracture to the tarsometatarsus. The images show masses in the gut and throat areas both containing small mammal bones. The bundle has granular inclusions which may be sand particles and the skull contains dehydrated brain matter.

MA1.27 – (DOAY005, 1.1983.159)
Identification – *Falco tinnunculus* (Kestrel)
Radiographic Report – XR/CT – Complete mummy bundle containing the skeletal remains of a kestrel and small granular inclusions. The legs appear flexed slightly with the claws splayed. The AP image shows that the legs are crossed. The tip of the beak is attached, but protrudes from the bandages. Eye matter is visible in the orbital cavity.

MA1.28 – (DOAY007, 1.1983.156)
Identification – *Falco naumanni* (Lesser Kestrel)
Radiographic Report – XR – Complete mummy containing the skeleton of a bird with the inclusion of the lower half of a crocodile skeleton lying along the breast. The bird skeleton displays what appears to be a defect to the spinal column of the bird and a compression to the top of the skull. There is a dense area around the neck and the tail area, and a mass in the gut. The outer layer of bandages is loose.

CT – Experiments were carried out with the threshold to attempt to gain a better insight into the bundle contents. These revealed that altering the attenuation using computer software can be used to visualise the remaining soft-tissue. This method was also used to show the crocodile skeleton more clearly. As the images were magnified to look at areas of interest it was noted that detail was reduced. This is shown well on reformations. Overall, this specimen is an excellent case study for the advantages and disadvantages of CT.

MA1.29 – (DOAY008, 1.1983.69a)
Identification – *Accipiter nisus* (Sparrowhawk) or *Accipiter brevipes* (Levant Sparrowhawk)
Radiographic Report – XR/CT – Incomplete mummy bundle containing skeletal material and dense granular pieces. The skull appears to belong to a different bird, the front of which has been lost. There are fractures and displacement of bones within the bundle.

MA1.30 – (DOAY010a, 1.1983.69c)
Identification – *Accipiter nisus* (Sparrowhawk) or *Accipiter brevipes* (Levant Sparrowhawk)
Radiographic Report – XR/CT – Mummy bundle containing the fully articulated skeleton of a mature bird. The skull has become separated from the body. One femur and wing show fractures and there is a small dense mass near the claws.

MA1.31 – (DOAY011, 1.1983.285.1)
Identification – *Falco tinnunculus* (Kestrel)
Radiographic Report – XR/CT – Complete mummy bundle containing the fully articulated skeleton of a bird. The bandages are only visible as a thin layer around the body. The tip of the beak has been lost post-mortem. There is evidence of a small mass in the gut area and dehydrated brain and eye matter is visible.

MA1.32 – (DOAY012, 1.1983.285.2)
Identification – *Falco tinnunculus* (Kestrel)

Radiographic Report – XR – Mummy contains the skeleton of a mature bird. The skull has been lost and one wing is fractured. There is a mass containing small bones in the gut area. The bandages appear on the images as a thin layer and the claws are uncovered.

CT – This specimen demonstrates the poor quality spatial resolution which CT provides opposed to X-ray. It is likely that the mass in the gut area is organ residue or stomach contents. Overall, CT adds little further information.

MA1.33 – (DOAY014, A.126.1968)
Identification – *Falco tinnunculus* (Kestrel)
Radiographic Report – XR/CT – Complete bird skeleton except for the loss of the front of the skull which has become separated from the body. There is a dense area near the tail. The images clearly show pseudo-pathology caused by superimposition of bones.

MA1.34 – (DOAY016, 1.1983.88)
Identification – Avian (Unidentifiable)
Radiographic Report – XR/CT – Incomplete mummy bundle containing half of a bird skeleton.

MA1.35 – (DOAY036)
Identification – *Falco tinnunculus* (Kestrel)
Radiographic Report – XR – The bundle is incomplete as the skull has become detached. The bundle contains two bird legs only complete with claws. There is a skull included with this, but separated from it and it is believed that it is unlikely to belong to the same animal as the bundle. The bandaging is thick in places and has been moulded into the shape of a bird mummy.

CT – The CT shows the presence of pinion feathers that have been used to reinforce the shape of the mummy bundle.

MA2.18 – (DOAY034, 1.1983.87)
Identification – Feline, juvenile
Radiographic Report – XR – Incomplete mummy bundle containing the skeleton of a juvenile cat. There is a break in the spinal column leading to separation and loss of vertebrae and some of the thorax. The body has been densely wrapped with only the bandages holding the two halves of the skeleton together. The exterior of the bundle is lavishly decorated with the addition of painted facial features. Brain matter is visible in the skull cavity. The tail appears to have been removed post-mortem, but prior to mummification.

CT – The CT of this specimen confirms the absence of certain skeletal components, however the question was raised of whether they had been lost post-mummification or whether they had never been in the bundle as there is no visible hole through which they could have been lost. The brain matter has settled to the rear of the skull. The CT clearly shows the growth plates on the lower limbs.

MA2.19 – (DOAY035, 91,890.00)
'Recorded from Beni Hasan'
Identification – Feline, mature
Radiographic Report – XR/CT – An unwrapped feline skull coated in a tar-like substance. The images show that the dentition is permanent and the root canals are closed indicating that the animal is mature. No vertebrae are visible.

MA2.20 – (?)
Identification – Feline
Radiographic Report – XR/CT – Complete mummy bundle in the form of a cat, but containing only one cat leg complete with paw.

MA2.21 – (skull)
Identification – Feline
Radiographic Report – XR/CT – Unwrapped cat skull with five articulated cervical vertebrae. The ears are folded in close to the skull and yellowish fur is visible. The teeth are exposed in areas.

MA4.12 – (DOAY009, 1.1983.69b)
Identification – *Lates niloticus* (Nile Perch)
Radiographic Report – Complete skeleton of a fish except for the front of the skull is missing. There is an area of dense material in the gut which is likely to have been ingested.

MA4.13 – (DOAY020, 1.1983.?)
Identification – *Lates niloticus* (Nile Perch)
Radiographic Report – XR/CT – Exquisitely modelled fish mummy containing the remains of a fish but positioned the wrong way round inside the bundle (the face is painted on at the tail end). A smaller fish is visible in the gut area. The whole bundle is wrapped in bandages and facial features have been modelled from linen on the exterior.

MA4.14 – (DOAY021, 10.1920.9)
'Graeco-Roman from Esneh fish cemetery'
Identification – *Lates niloticus* (Nile Perch)
Radiographic Report – XR/CT – Complete fish mummy containing the skeleton of a fish. There is a fracture in the spinal column which has led to some displacement, however the rest of the skeleton is fully articulated. Dehydrated brain matter is visible in the skull cavity.

MA5.9 – (DOAY025, 42.12)
'Recorded from Hawara and of the Graeco-Roman period'
Identification – *Crocodylus niloticus* (Nile crocodile), juvenile
Radiographic Report – XR/CT – Unwrapped crocodile skull fragments presented in a box along with a dense mass containing granular fragments. The bones are not articulated or compressed in any way. The box contains the remains of more than one animal, however it is not possible to say how many.

MA5.10 – (DOAY031, 58.1911f)
'part of a group of eight crocodiles from Hawara'
Identification – *Crocodylus niloticus* (Nile crocodile), juvenile

Radiographic Report – XR/CT – Complete mummy bundle containing the skeletal remains of the upper half of a crocodile skeleton severed before the pelvis. There is evidence to suggest curvature of the spine. The body shape of the crocodile has been created from reeds laid longitudinally and bound with bandages. The mummy has got individually modelled limbs and a bent tail.

MA5.11 – (DOAY010b)
Identification – *Crocodylus niloticus* (Nile crocodile), juvenile
Radiographic Report – XR/CT – Complete mummy bundle which contains the upper half of a crocodile skeleton including the skull, cervical spine and upper limbs. The lower half of the mummy bundle appears to be formed from fabric.

MA5.12 – (DOAY022, 1.1983.95)
'Recorded from Memphis'
Identification – *Crocodylus niloticus* (Nile crocodile), juvenile
Radiographic Report – XR/CT – Complete mummy of a juvenile crocodile. The skull has become separated and brain and orbital matter are visible. The mummy is wrapped in loose bandages. Two fractures in the spinal column close to the pelvis are visible on the images. The limbs are in a natural position which is in contrast to the tightly wrapped mummies where the limbs are generally folded in close to the body.

MA5.13 – (DOAY023a, 1.1983.77)
'one of three mummified crocodiles from near Kom Ombo'
Identification – *Crocodylus niloticus* (Nile crocodile), juvenile
Radiographic Report – XR/CT – Complete mummified crocodile skeleton showing a layer of tar-like substance on the exterior and no bandages. Granular inclusions are visible attached to the exterior presumably to the sticky substance. Brain matter is visible in the skull cavity and there is a fracture visible in the spinal column.

MA5.14 – (DOAY023b, 1.1983.77)
'one of three mummified crocodiles from near Kom Ombo'
Identification – *Crocodylus niloticus* (Nile crocodile), juvenile
Radiographic Report – XR/CT – Complete skeleton of a crocodile except for the loss of a front limb. There is a fracture visible in the spinal column, and the skull and a section of the rib cage have become separated form the body. The exterior of the body has been coated in a tar-like substance which has attracted granular pieces. Brain matter is visible in the skull cavity.

MA5.15 – (DOAY023c, 1.1983.77)
'one of three mummified crocodiles from near Kom Ombo'
Identification – *Crocodylus niloticus* (Nile crocodile), juvenile
Radiographic Report – XR/CT – Skeleton of a juvenile crocodile missing the end of the tail. The specimen is not wrapped and the exterior of the animal has been coated in a tar-like substance which has attracted granular material. Brain matter is visible in the skull cavity.

MA5.16 – (DOAY024)
Identification – *Crocodylus niloticus* (Nile crocodile), juvenile
Radiographic Report – XR/CT – Incomplete crocodile skeleton with the end of the tail missing. The animal has been wrapped in loose bandages. The front limbs are folded in close to the body and the hind limbs are flexed out.

MA6.5 – (DOAY018, A.128.1968)
Identification – Snake, juvenile
Radiographic Report – XR/CT – Complete mummy bundle containing the skeleton of a juvenile snake. The body has become separated into segments as it decomposed. The bandaging is thick and the mummy appears to contain granular material and debris.

MA6.6 – (DOAY017, 1.1983.88)
Identification – *Gerbillus* (Gerbil)
Radiographic Report – XR/CT – Complete mummy bundle containing the skeletal remains of a mature gerbil in a layer of thick bandages.

MA6.7 – (DOAY019, 1.1983.292)
Identification – Egg
Radiographic Report – XR/CT – Complete mummy bundle containing an egg thought to be that of a chicken. The specimen is not wrapped. The egg has cracked and has become compressed over time.

MA7.18 – (DOAY001, 1.1983.288)
'Mummified ibis'
Identification – Pseudo-mummy
Radiographic Report – XR – Complete mummy containing bone and egg-shell fragments mixed with granular material.

CT – CT shows that the objects within the bundle have settled to the base with an air-filled void above. Wrappings laid in lines are visible. There is a small collection of vertebrae which could be from the vertebral column of a snake.

MA7.19 – (DOAY006)
Identification – Pseudo-bird
Radiographic Report – XR – Complete mummy of a bird modelled in a standing position containing no skeletal material and no visible anomalies. The outer layers of bandages are clearly visible on the head. One false leg has become separated from the body.

CT – The CT shows that the eyes are made from fabric with an outer level made of a high density material.

MA7.20 – (DOAY013)
Identification – Pseudo-bird
Radiographic Report – XR – Complete mummy bundle containing no skeletal material. There are two visible

anomalies within the bundle which are either semi-calcified eggs or bundles of fabric containing granular material. The mummy wrappings are 25mm thick at the deepest point.

CT – Shows that the bandages have been applied in three distinct layers. The CT has added a lot of information about the wrappings.

MA7.21 – (DOAY015, 15.88)
Identification – Pseudo-bird
Radiographic Report – XR/CT – Complete bundle containing no identifiable skeletal material. The mummy bundle appears to have been formed from egg-shell fragments, gunge and small unidentifiable bone/feather fragments. The bundle has become damaged in one area with the contents protruding out. The wrappings appear to be thin and patchy.

MA7.22 – (DOAY026, 58.1911a)
'part of a group of eight crocodiles from Hawara'
Identification – Pseudo-crocodile
Radiographic Report – XR – Complete mummy bundle containing no skeletal material. There is a large tooth-shaped object visible on the images possibly made of stone or metal. Two metal pins are also visible, but are thought to be modern additions used to fix the mummy to a presentation board. The body shape of the crocodile has been created from reeds laid longitudinally and bound with thin thread.

CT – The CT shows the presence of reeds within the bandages really clearly with the air trapped within them.

MA7.23 – (DOAY027, 58.1911b)
'part of a group of eight crocodiles from Hawara'
Identification – Pseudo-crocodile
Radiographic Report – XR/CT – Complete mummy bundle containing no skeletal material. The mummy contains egg shell fragments and two metal pins which are likely to be modern and relate to the method of presentation at the museum. The body shape of the crocodile has been created from reeds laid longitudinally and bound with thin thread.

MA7.24 – (DOAY028, 58.1911c)
'part of a group of eight crocodiles from Hawara'
Identification – Pseudo-crocodile
Radiographic Report – XR/CT – Complete mummy bundle containing no skeletal material. There are egg shell fragments and two metallic objects (possibly metal film or gilding) within the bundle. The body shape of the crocodile has been created from reeds laid longitudinally and bound with thin thread.

MA7.25 – (DOAY029, 58.1911d)
'part of a group of eight crocodiles from Hawara'
Identification – Pseudo-crocodile
Radiographic Report – XR – Complete mummy bundle containing no skeletal material. There are egg shell fragments and two metallic objects (possibly metal film or gilding) within the bundle. The body shape of the crocodile has been created from reeds laid longitudinally and bound with thin thread.

CT – The dense object within the bundle has an extremely high attenuation (1253.6) which is similar to that shown by cortical bone or metal. A 'region of interest' was created using computer software to analyse the density of this area.

MA7.26 – (DOAY030, 58.1911e)
'part of a group of eight crocodiles from Hawara'
Identification – Pseudo-crocodile
Radiographic Report – XR/CT – Complete mummy bundle containing no skeletal material. There is a tooth with an open root within the bundle which is thought to belong to a pig, jackal or hyena. The body shape of the crocodile has been created from reeds laid longitudinally and bound with thin thread.

MA7.27 – (DOAY033, 58.1911h)
'part of a group of eight crocodiles from Hawara'
Identification – Pseudo-crocodile
Radiographic Report – XR/CT – Complete mummy bundle containing no skeletal material. There is a human middle phalanx bone incorporated within the mummy bundle along with two metal pins. The body shape of the crocodile has been created from reeds laid longitudinally and bound with thin thread.

MA7.28 – (DOAY032, 58.1911g)
'part of a group of eight crocodiles from Hawara'
Identification – Pseudo-crocodile
Radiographic Report – XR/CT – Complete mummy bundle containing no skeletal material. There are two metal pins and granular material visible on the images. The body shape of the crocodile has been formed from a mud-like substance with tail, legs and eyes modelled individually. The tail is fixed at a ninety degree angle to the body. The body has been covered with reeds before being covered with a tar-like substance which has attracted granular particles.

MA8.2
Identification – Pseudo-baby
Radiographic Report – XR/CT – Unusual object modelled to resemble a human doll complete with individual limbs and facial features. One leg has been shortened by the loss of the end of the limb. The stomach area is concave. The images show that the specimen is devoid of human or animal material and has been modelled from a mud-like material. A central stick has been inserted into the skull area to provide support and a plaster face has been added. The age of the specimen is unknown, however it is possible that the object is a modern forgery intended to look like a human baby or doll.

MA8.3
Identification – Pseudo-baby
Radiographic Report – XR/CT – Unusual object resembling a mummified bird in form (conical shape) with the addition of a plaster face mask displaying human

facial features. The mummy is devoid of human or animal material and appears to have been made from a mud-like substance wrapped in bandages. The object is presented in a wooden display coffin with a glass lid.

MANCHESTER MUSEUM
Forty-five mummies
Contact Details:
The Manchester Museum, University of Manchester, Oxford Road, Manchester M13 9PT
Tel. 0161 275 8772
Fax. 0161 275 2676
Email. museum@manchester.ac.uk

There are no CT results for the Manchester collection as the mummies were not CT scanned due to restricted access to the collection at the time the study was being carried out.

MA1.10 – (MM3, 6035)
'From Aswan. Donated by J. And A. Williams'
Identification – *Falco tinnunculus* (Kestrel)
Radiographic Report – XR – Complete mummy bundle containing the skeleton of a mature kestrel. There is an area of opacity in the region of the pelvis which may represent preserved organs. The spine shows a level of distortion which probably occurred during the mummification process. The bird has been tightly wrapped with the densest area of bandages in the neck area. The metallic disc visible on the images is a museum label.

MA1.11 – (MM17, 9371)
'Purchased in 1934'
Identification – *Accipiter nisus* (Sparrowhawk), mature
Radiographic Report – XR – Complete mummy bundle containing the skeletal of a mature bird. There is a loss of skeletal integrity in the spinal column with some displacement of vertebrae. There are granular fragments near the legs. The specimen is tightly bound in brown linen bandages. The bird has long meta-carpels and phalanges indicative of the sparrowhawk.

MA1.12 – (MM18, 9428, 9248)
'Donated by Mrs E. Thomas, 1936'
Identification – *Falco tinnunculus* (Kestrel), mature
Radiographic Report – XR – Complete mummy bundle containing the fully articulated skeleton of a kestrel, complete except for the tip of the beak which is missing. The bird is tightly bound in brown bandages.

MA1.13 – (MM19, 9429)
'Donated by Mrs E. Thomas, 1936'
Identification – *Falco tinnunculus* (Kestrel), mature
Radiographic Report – XR – Complete mummy bundle containing the remains of a kestrel with the tip of the beak exposed from the bandages. The bird has been tightly wrapped in bandages.

MA1.14 – (MM21, 11294, Robinow 70)
'Late Period mummy from the Robinow Collection, 1959'
Identification – *Falco tinnunculus* (Kestrel)
Radiographic Report – XR – Complete mummy bundle containing the fully articulated skeleton of a kestrel. The limbs are flexed towards the the body in an unnatural position. Harris Lines are visible in the humerus suggesting that the bird suffered from a period of ill-health from which it recovered. The bird is unwrapped.

MA1.15 – (MM22, 11295, 22944, Robinow 69)
'Late Period mummy from the Robinow Collection, 1959'
Identification – *Falco tinnunculus* (Kestrel), mature
Radiographic Report – XR – Complete mummy bundle containing the fully articulated skeleton of a bird. The images show the presence of air trapped within the feathers shown as radiolucent areas. There is a round opacity in the region of the thoraco-abdominal junction which could represent organs. There is an opaque mass near the joint of the tibiotarsus and the tarsometatarsus which could represent metal or a material of a high atomic number. The mummy is wrapped in dark brown bandages with the beak and one orbital socket exposed.

MA1.16 – (MM23, 1971.21)
'Bird mummy from North Saqqara, XXVI-XXX dynasty. From the Egypt Exploration Society excavations (Ex. No. H5-2587, Inventory 4864). Come to Manchester via the British Museum in 1971'
Identification – *Falco tinnunculus* (Kestrel), mature
Radiographic Report – XR – Complete mummy containing the semi-articulated skeleton of a mature kestrel. The bundle has been fractured transversely in four places which has led to the fracture of one tibiotarsus and the loss of the beak and right eye. There is evidence for a metallic inclusion between the bird and the wrappings and resinous deposits appear as speckled areas on the images. The mummy has been wrapped in brown bandages with an ornamental pattern on the chest area. Facial features have been painted on the exterior in black and red.

MA1.17 – (MM27, 1981.576)
Identification – *Accipiter nisus* (Sparrowhawk)
Radiographic Report – XR – Complete mummy bundle containing the fully articulated skeleton of a hawk. There is a layer of air visible trapped between the feathers. Metal pins have been used to secure the head and feet during a modern conservation attempt. The skeleton is considerably smaller than the mummy bundle itself. Interestingly the tarsometatarsus is thicker and longer in this specimen than in others.

MA1.18 – (MM29, ?)
'Attached label reads 'Mummy of a Sacred Hawk. This having been cut open, shows the bird and the many wrappings of linen. The same species of hawk (Falco tinnunculoides) is still found today in Egypt. From Thebes c200BC'
Identification – *Falco tinnunculus* (Kestrel)
Radiographic Report – XR – This specimen has been dissected at some point to investigate the contents and is presented in its dissected state. The skeleton of the bird

and the feathers are visible within the layers of bandages. The dissection process has caused considerable disruption to the articulation of the skeleton. The feet are missing and it is thought that the tibiotarsus was amputated at the time of dissection due to the resulting margins of the bones. The cervical spine is connected to the skull, but this is disarticulated from the rest of the skeleton. The bandages are brown linen.

MA1.19 – (MM31, 4295)
Identification – *Falco tinnunculus* (Kestrel)
Radiographic Report – XR – Complete bundle containing the incomplete skeleton of a hawk. There is an incomplete lower limb including the distal portion of the femur and foot in a claw-like position, part of the pelvis, humerus and scapula. There are areas of density thought to be sand or resin. The mummy bundle has been wrapped in a geometric pattern with a modelled head made from a dense material.

MA1.20 – (MM40, 6098)
'Roman mummy from Abydos Ibis Cemetery. From the Egypt Exploration Fund, 1913-1914'
Identification – *Threskiornis aethiopica* (Sacred Ibis)
Radiographic Report – XR – Complete mummy bundle containing the complete skeleton of an ibis. The images are relatively poor as the ibis was packed for transportation in a box surrounded by vermiculite which appears as a snowy texture. There is evidence of disruption to the spinal column and the skull has been detached and placed between the legs of the bird. A radio-opacity at the base suggests that the resin has set in different ways when the cartonnage was placed in a standing position.

The mummy bundle is wrapped in dark and light brown bandages arranged in a diagonal pattern with the addition of a false head modelled to represent Thoth and Atef-crown supported behind the head.

MA1.21 – (MM41, 11501)
'Late Period mummy from Saqqara. From the Egypt Exploration Society via the British Museum, 1969'
Identification – *Threskiornis aethiopica* (Sacred Ibis)
Radiographic Report – XR – Complete mummy imaged in three sections because of its size with overlap. The skeleton is disarticulated possibly because the contents have moved over time within the bundle. The bones appear as a complete jumble of bones with no correlation to their anatomical position and there is no evidence of the presence of a skull or claws. The mummy bundle is decorated with diagonal bandaging decorated with Thoth seated on a throne and surmounted by the Atef-crown.

MA2.8 – (MM13, 6842)
'Donated by M.E. Robinow in 1922'
Identification – Feline, juvenile (*Felis silvestris libyca*)
Radiographic Report – XR – Complete mummy bundle containing the skeletal remains of a juvenile cat. The limbs are not clearly visible on the images, but are present. There is a mass of small pellets in the throat area and the skull shows signs of crushing. The mummy is wrapped in brown linen bandages with the addition of false linen ears and facial features painted in black and red.

MA2.9 – (MM14, 1977.1162)
'From the National Museum of Wales, Cardiff. March 1898'
Identification – Feline, juvenile
Radiographic Report – XR – Complete mummy bundle containing the skeleton of an immature cat evidenced from the presence of unfused epipyses in a number of the long bones. The mummy has been imaged in two halves with an overlap because of the size (425mm in length). There is a loss of skeletal integrity in the spinal column leading to the loss of a number of the cervical vertebrae. There is a fracture through the right radius and ulna and a dislocation at the elbow. The skull has been crushed. The mummy has been wrapped in brown linen bandages and the speckled appearance on the images could represent sand or resin on the skin. A metal pin close to the lower limbs could represent a conservation attempt to stabilise the mummy.

MA2.10 – (MM15, 11123, Robinow 68)
'From the Robinow Collection, 1959'
Identification – Feline, juvenile (*Felis silvestris libyca*)
Radiographic Report – XR – Skull only dislocated at the cranio-cervical junction. There is evidence of brain matter and small teeth. The skull is unwrapped and there is a layer of resin and sand over the skin.

MA2.11 – (MM16, 12015, 22948)
'Middle Kingdom mummy from Beni Hasan'
Identification – Feline, mature (*Felis silvestris libyca*)
Radiographic Report – XR – Skull with five fully articulated cervical vertebrae attached. There is evidence of brain matter and teeth in-situ. Large canine teeth are visible in the upper jaw. There is a metallic disc near the mandibular condyle.

MA2.12 – (MM25, 12012, 22945, EG9)
'Late Period mummy'
Identification – Feline, juvenile (*Felis silvestris libyca*)
Radiographic Report – XR – Complete mummy bundle containing the fully articulated skeleton of a juvenile cat. The cat has been bound tightly with the lower limbs folded up and the upper limbs folded down with the tail between the legs. There is a fracture through the mid thoracic spine leading to displacement of the vertebrae and there is evidence of damage to the maxilla. The animal has long metacarpals and metatarsals.

MA2.13 – (MM28, A957)
Identification – Feline, juvenile (*Felis manni*culata)
Radiographic Report – XR – This is an unwrapped cat mummy presented in a glass lidded box. The cat has visible yellowish fur. It is not a complete skeleton – there is a skull, disrupted cervical spine, incomplete thoracic spine, two complete upper limbs and four lumbar vertebrae. It is difficult to assess age as it is impossible to confirm whether there is primary of erupting secondary

dentition; however, the size of the cat suggests that it is immature.

MA2.14 – (MM34, 12016)
Identification – Feline, mature
Radiographic Report – XR – Unwrapped skull of a mature cat with five articulated cervical vertebrae. Brain matter is visible in the skull cavity and resin is noticeable on the skin. The dentition is all in-situ.

MA2.15 – (MM36, ?)
'Attached label reads March 1898'
Identification – Feline, mature
Radiographic Report – XR – Complete mummy bundle containing the semi-articulated skeleton of a mature cat. There is considerable disruption in the thoraco-lumbar region and the upper limbs are disarticulated at the elbow. Some areas of the skeleton are hard to view on the images due to superimposition. There is an opacity in the torso region which could be resin or a package. The spine has been bent during the mummification process and there are no visible signs that the neck has been broken. There are loose bandages holding the two halves of the bundle together.

MA3.2 – (MM12, 563)
'Mammal mummy dating to the XVIII – XIX dynasty from the site of Gurob. Collected by Petrie and donated to the museum by Jesse Haworth in 1906'
Identification – Canine, juvenile
Radiographic Report – XR – Mummy of an immature puppy which has lost the skull and the proximal cervical vertebrae. Only one upper limb is visible on the images. The skeleton is tightly bound with the rear limbs folded upwards, the upper limbs folded down and the tail between the legs. The mummy is wrapped in brown linen bandages. The speckled texture visible on the images could represent sand or resin on the fur.

MA4.11 – (MM30, ?)
'From E. J. Woodward, School and Museum Resource Service, Wakefield, 1980'
Identification – Fish (Possibly sp. *Clarias* – Catfish)
Radiographic Report – XR – Complete mummy bundle containing the skeleton of a fish. There is a fracture in the mid-section of the mummy with some loss of skeletal integrity. The vertebrae appear as X-shaped bones on the images. There are a number of granular inclusions which may be inside the fish or within the bandages. The dentary appears to be fairly thick and rounded with barbells and there is an attractive pattern to the cartilaginous structure of the head. The mummy is fairly fragile and is composed of reeds and loose bandages.

MA5.1 – (MM1, 100/1, EA19/1)
'Roman period mummy of a crocodile from the British School of Archaeology 1911. The object was previously kept at the University of Aberdeen Anthropological Museum, Mariscall Collection'
Identification – *Crocodylus niloticus* (Nile crocodile), juvenile
Radiographic Report – XR – Excellently preserved mummy of a juvenile Nile crocodile which has lost the end of the tail. There is a fracture through the lower thorax. The upper limbs are folded close to the body and the lower limbs are in a natural position. The crocodile is totally unwrapped.

MA5.2 – (MM2, 100/2, EA19/2)
'Roman period mummy of a crocodile found at Hawara from the British School of Archaeology 1911. The object was previously kept at the University of Aberdeen Anthropological Museum, Mariscall Collection'
Identification – *Crocodylus niloticus* (Nile crocodile), juvenile
Radiographic Report – XR – Mummy of a juvenile crocodile with the head unwrapped and the body covered in loose light brown bandages. The skeleton is complete except for the end of the tail which is missing. There is a fracture in the proximal tail. The images show an overlapping of the limbs over the spinal column which causes loss of definition.

MA5.3 – (MM5.3, 7892)
'Possibly Roman. Donated by W. Sharp Ogden in 1925'
Identification – *Crocodylus niloticus* (Nile crocodile), juvenile
Radiographic Report – XR – Complete mummy containing the skeleton of a juvenile crocodile with the end of the tail missing. There is an area of radio-lucence with radio-dense margins possibly metallic paint on a reed or thin board. There are high density areas near the tail and head possibly resin or sand. The mummy has been wrapped in dark and light brown linen bandages in a rectangular pattern with evidence of reeds within the bandages.

MA5.4 – (MM9, 1772)
'Donated by M.E. Robinow between 1895 and 1896'
Identification – *Crocodylus niloticus* (Nile crocodile), mature
Radiographic Report – XR – This mummy was studied in five sections with overlap due to its large size (1697mm in length). The mummy is complete and contains the skeleton of a crocodile with an estimated age at death of over five years. The spinal column appears to be intact with no fractures. There is a ball-shaped mass in the stomach with dense opacities (possibly stones) and granular fragments. There is a linear transparency in the occipital region and disruption of what is thought to be the peri-orbital bone. There is evidence of fractures with fragments of bone displaced anterially which may be the mandible or maxilla. There is evidence of post-mortem tooth loss. The damage to the skull is thought to be post-mortem; however it could represent the cause of death.

MA5.5 – (MM10, 12008, 22941, EG95)
'Mummy thought to date to the Ptolemaic period. Donated to the museum by M.E. Robinow'
Identification – *Crocodylus niloticus* (Nile crocodile)
Radiographic Report – XR – The mummy is wrapped in three sections of diagonal bandages with a straight line on

top. The position of the eyes has been marked with protruberances made from linen. There is evidence of reeds within the bandages. The images show that the mummy bundle contains the skulls of three crocodiles positioned in a line. There is an area of interest to either side of the largest skulls which may represent two smaller crocodiles, one lying on either side, or the remains of a snake.

MA5.6 – (MM11, 12010, 22943)
'Late Period mummy'
Identification – *Crocodylus niloticus* (Nile crocodile), juvenile
Radiographic Report – XR – Complete crocodile mummy containing the remains of a juvenile crocodile with the upper limbs folded close to the body and the lower limbs in a natural position. There is a dense mass in the stomach area. The bundle has been fully compressed at some time and the tail has been curved round. The entire bundle has a thin covering of light brown linen bandages bound with thread.

MA5.7 – (MM32, Robinow 72)
'From the Robinow Collection'
Identification – *Crocodylus niloticus* (Nile crocodile), juvenile
Radiographic Report – XR – Complete mummy bundle containing the distal half of a crocodile skeleton modelled into the form of a larger animal. The skeletal remains include the distal skeleton and lumbar spine of a juvenile crocodile with the top half obscured by a dense mass. This mass may contain small bones and there is evidence of resin and sand on the skin of the crocodile. There are reeds within the bandages and a geometric pattern has been created on the exterior of the bundle.

MA5.8 – (MM38, Sotherby's lot number 134944)
Purchased from Sotherby's in 1984'
Identification – *Crocodylus niloticus* (Nile crocodile), juvenile
Radiographic Report – XR – Complete mummy bundle containing the semi-articulated skeleton of a crocodile. There is a fracture through the lumbar spine and the body has separated into two halves, but there appears to be no loss of skeletal material. The loose brown bandages appear radio-lucent on the images. There are three granular pieces in the lower abdomen.

MA6.1 – (MM8, 3005)
'Roman crocodile mummy from Hawara. Donated to the museum by Jesse Haworth between 1887 and 1888'
Identification – Snake
Radiographic Report – XR – The mummy is formed in the shape of a crocodile from reeds wrapped in light brown bandages. It contains a mass of tiny snake vertebrae in a coiled position. The metallic disc on the image is a museum label.

MA6.2 – (MM42, 6033a/b)
'Bought at Aswan and donated by J. and A. Williams'
Identification – Small mammal (*Crocidura flavescens olivieri* or *Suncus murinus*)
Radiographic Report – XR – Complete mummy bundle containing the skeleton of a small mammal. The animal has prominent incisors. Attached to the back of the bundle is a smaller package with radio-lucent contents which is thought to represent a food package for the animal.

MA6.3 – (MM43, 1195)
'XXVI dynasty mummy from Hibeh. From the Egypt Exploration Fund in 1903'
Identification – Snake, juvenile
Radiographical Report – XR – Complete mummy bundle containing the semi-articulated remains of a snake. The skeleton is surrounded by a radio-opaque substance probably containing sand or mud. Decomposition has lead to the fragmentary nature of the contents. No fabric is visible and the mummy bundle is packed in a crudely made wooden box the lid of which is decorated with a carved snake.

MA7.6 – (MM4, 6293)
'Cat mummy dating from the XXII dynasty from Beni Hasan. Collected by Prof. Garstang and brought from the Liverpool Institute of Archaeology in 1920'
Identification – Pseudo-cat
Radiographic Report – XR – Complete mummy bundle in reasonable condition containing no skeletal material. There are a number of radio-patches within the bundle which may be resin or sand. The bandages form a pattern of concentric squares with false ears made of textiles. The wrappings are loose and there is a layer of trapped air between them and the object itself.

MA7.7 – (MM5, 5373c)
'Roman crocodile mummy from Hawara. Collected by Petrie and donated to Manchester between 1910 and 1911'
Identification – Pseudo-crocodile
Radiographic Report – XR – Complete mummy containing no skeletal material. The images show fragments of egg shell. The form of the crocodile has been created from reeds laid longitudinally then bound with thread. The head area appears to be more radio-dense than the rest of the body possibly due to the application of resin.

MA7.8 – (MM6, 5373d)
'Roman crocodile from Hawara. Collected by Petrie and donated to Manchester between 1910 and 1911'
Identification – Pseudo-crocodile
Radiographic Report – XR – Complete mummy containing no skeletal material. The images show fragments of egg shell in the head and neck area. There is an area of radio-opacity in the tail region thought to represent metallic paint.

MA7.9 – (MM20, 11293, Robinow 71)
'Late Period mummy from the Robinow Collection, 1959'
Identification – Pseudo-hawk
Radiographic Report – XR – Complete mummy bundle modelled in the shape of a bird. The mummy has been

tightly wrapped in brown bandages with evidence of reeds inside. There is a gilded head and breast cover which appears as an opaque area on the images. The bundle is completely empty of animal bones.

MA7.10 – (MM24, 6034)
'Mummy purchased at Aswan and donated to the museum by J. and A. Williams'
Identification – Pseudo-mammal
Radiographic Report – XR – Complete mummy bundle containing no skeletal material. There are dense areas in the mid-section of an uncertain material and not representing any real structure. The mummy has been tightly bound in dark brown linen bandages. It is possibly that the mummy was formed to resemble a snake; however, there are no animal bones inside.

MA7.11 – (MM26, 12013, 22946, EG8)
Identification – Pseudo-mammal
Radiographic Report – XR – Complete mummy bundle containing no skeletal material. The body shape has been formed from a dense granular material before being wrapped. There are a number of air pockets visible on the images in the mid-section of the mummy.

MA7.12 – (MM33, Sotherby's lot number 134944)
'Purchased at Sotherby's, 1984'
Identification – Pseudo-feline
Radiographic Report – XR – Complete mummy bundle containing no animal bones. There is an area of radio density in the distal aspect of the bundle, but not dense enough to be bones. The bandages form a geometric design.

MA7.13 – (MM35, 9849, EA16/2)
Identification – Pseudo-feline
Radiographic Report – XR – Complete mummy bundle containing no animal bones. There is a fracture through the mid-section of the bundle containing trapped air. One false ear is missing. The body shape has been created from a dense material containing what could be sand and resin.

MA7.14 – (MM37, ?EA16/2)
Identification – Pseudo-mammal
Radiographic Report – XR – Complete mummy bundle containing a mass of tiny unidentifiable bones with visible growth plates which have fallen to the bottom of the holder. There is a mass which could be a skull in the mid-section. The bundle is unusual in that it resembles a sock-like holder made of brown linen with individually wrapped limbs bound with thread. The bundle resembles a child's toy but is likely to be a young or maybe even stillborn animal.

MA7.15 – (MM44, 6032)
'Bought at Aswan and donated by J. and A. Williams'
Identification – Pseudo-rodent
Radiographic Report – XR – Complete light brown oval shaped mummy bundle containing unidentifiable remains. It is likely to contain what appears to be a bird modelled from a radio-opaque substance.

MA7.16 – (MM45, 11296, Robinow 67)
'Possibly Ptolemaic mummy from the Robinow Collection, 1959'
Identification – Pseudo-ibis
Radiographic Report – XR – Complete mummy bundle containing no skeletal material. The bandaging is arranged diagonally with evidence of reeds laid longitudinally inside.

MA7.17 – (MM46, 5373b)
'Roman mummy from Hawara. Collected by Petrie, 1910-11'
Identification – Pseudo-crocodile
Radiographic Report – XR – Complete mummy bundle containing no skeletal material. There is a dense object lying between the reeds and the outer packaging, and a dense granular area. The end of the tail had become detached but is present.

STOCKPORT METROPOLITAN BOROUGH COUNCIL

Four Specimens
Contact Details:
Heritage Services, Stockport Metropolitan Borough Council, Woodbank Hall, Woodbank Park, Stockport SK1 4JR
Tel. 0161 474 4452

MA1.23 – (LX19a)
Identification – cf. Hawk
Radiographic Report – XR/CT – Tightly bound bird mummy containing the complete and fully articulated skeleton of a hawk. The head is folded close to the breastbone in an unnatural position. The lateral images show a dense area at the base of the mummy bundle which may be resin which has settled in this position following the mummification process.

MA2.16 – (LX18.1)
Identification – Feline
Radiographic Report – XR/CT – Unwrapped feline skull with no attached cervical vertebrae. The dentition is visible in-situ.

MA2.17 – (LX18.2)
Identification – Feline
Radiographic Report – XR/CT – Unwrapped feline skull with five articulated cervical vertebrae.

MA6.4 – (STOPM.1992.432)
'Rat mummy recovered from a chimney in Stockport'
Identification – Rat
Radiographic Report – XR/CT – Complete unwrapped mummy of a rat presented in a cardboard box. The skeleton is fully articulated with the tail folded along the animal's spine. It is thought that this is a modern mummy.

MISCELLANEOUS

One specimen

MA1.22 – (Rosalie's Bird)
'Bird mummy loaned to Rosalie David by a private collector for research purposes'
Identification – Bird (cf. Hawk or cf. Falcon)
Radiographic Report – XR/CT – Excellently preserved mummy bundle containing the complete and fully articulated skeleton of a bird. The very tip of the beak has been lost probably because it protrudes from the bandages. The wings are flexed closely in to the body and the limbs are outstretched.

BIBLIOGRAPHY

ADAMS, J.E. and ALSOP, C.W. 2008. 'Imaging in Egyptian mummies' in R. David (ed) Egyptian Mummies and Modern Science, Cambridge University Press (New York).

AMBROSE, J. and HOUNSFIELD, G.N. 1973. 'Computerized transverse axial tomography', British Journal of Radiology 46 (542): 148-9.

ARMITAGE, P.L. and CLUTTON-BROCK, J. 1981. 'A radiological investigation into the mummification of cats from ancient Egypt', Journal of Archaeological Science 8: 185-196.

AUFDERHEIDE, A.C. 2003. The Scientific Study of Mummies, Cambridge University Press (Cambridge).

BRIER, B. 1996. Egyptian Mummies, Michael O'Mara Books Ltd (London).

BRIER, B. and BENNETT, M.V.L. 1979. 'Autopsies on Fish Mummies', Journal of Egyptian Archaeology 65: 128-133.

CLUTTON-BROCK, J. 1974. 'The Buhen Horse', Journal of Archaeological Science 1: 89-100.

D'AURIA, S.H. and MARX, H. 1988. 'Three-Dimensional CT Reconstructions of an Ancient Human Egyptian Mummy', American Journal of Radiology 150: 147-149.

DAVID, A.R. 1979. The Manchester Museum Mummy Project: Multidisciplinary Research on Ancient Egyptian Mummified Remains, Manchester Museum (Manchester).

DAVID, A.E. 1986. 'The Conservation of Mummified Egyptian Remains' in A.R. David (ed) Science in Egyptology, Manchester University Press (Manchester).

DAVIES, V. and FRIEDMAN R.F. 1998. Egypt, British Museum Press (London).

DODSON, A. 1995. 'Of Bulls and Princes: the early years of the Serapeum at Sakkara', KMT 6: 18-32 (San Francisco).

DODSON, A. 2005. 'Bull Cults' in S. Ikram (ed) Divine Creatures: Animal Mummies in Ancient Egypt, The American University in Cairo Press (Cairo) 72-105.

EMERY, W.B. 1965. 'Preliminary Report on the Excavations at North Saqqara, 1964-65', Journal of Egyptian Archaeology 51: 3-8.

EMERY, W.B. 1968. 'Preliminary Report on the Excavations at North Saqqara, 1968', Journal of Egyptian Archaeology 55: 31-35.

EMERY, W.B. 1970. 'Preliminary Report on the Excavations at North Saqqara, 1968-9', Journal of Egyptian Archaeology 56: 5-11.

EMERY, W.B. 1971. 'Preliminary Report on the Excavations at North Saqqara, 1969-70', Journal of Egyptian Archaeology 57: 3-13.

FALKE, T.H.M. 1997. 'Radiology of Ancient Egyptian Mummified Animals' in J. Van Dijk (ed) Essays on Ancient Egypt in Honour of Herman Te Velde, Styx Publications (Groningen).

FAULKNER, R.O. 2000. The Ancient Egyptian Book of the Dead, British Museum Press (London).

GABRA, S. *et al.* 1945. 'Rapport sur les fouilles d'Hermopollis Ouest Touna el-Gebel, CdÉ, 10 (104): 237-244.

GARNER, R.C. 1979. 'Experimental Mummification of Rats' in A.R. David (ed) Science in Egyptology, Manchester University Press (Manchester) 11-16.

GERMOND, P. 2001. An Egyptian Bestiary. Thames and Hudson (London).

HOFFMAN, H., W.E. TORRES, *et al.* 2002. 'Paleoradiology: Advanced CT in the Evaluation of Nine Egyptian Mummies.' Radiographics 22 (2): 377-85.

HOULIHAN, P.F. 1996. The Animal World of the Pharoahs, Thames and Hudson (New York).

HOUNSFIELD, G.N. 1973. 'Computerized transverse axial scanning (tomography)', British Journal of Radiology 46 (552): 1016-22.

HOUNSFIELD, G.N. 1976. 'Picture quality of computed tomography', American Journal of Roentgenology 127 (1): 3-9.

http://animaldiversity.ummz.umich.edu/site/resources/Grzimek_fish/Siluriformes/Clarias_gariepinus – website consulted on 25th June 2008 at 22:46.

http://animaldiversity.ummz.umich.edu/site/resources/Grzimek_fish/Percoidei/PercoideiIII/Lates_niloticus – website consulted on 25th June 2008 at 22:51.

http://museum.man.ac.uk/collections/egyptology/egyptology.htm – website consulted on 13th May 2006 at 21:19.

IKRAM, S. and ISKANDER N. 2002. Non-Human Mummies, The Supreme Council of Antiquities (Cairo).

IKRAM, S. 2004a. Beloved Beasts: Animal Mummies from Ancient Egypt, The Supreme Council of Antiquities (Cairo).

IKRAM, S. 2004b. A Zoo for Eternity: Animal Mummies from the Cairo Museum, The Supreme Council of Antiquities (Cairo).

IKRAM, S. (ed). 2005. Divine Creatures: Animal Mummies in Ancient Egypt, American University in Cairo Press (Cairo).

IKRAM, S. 2005. 'Manufacturing Divinity' in S. Ikram (ed) Divine Creatures: Animal Mummies in Ancient Egypt, American University in Cairo Press (Cairo) 16-43.

JACQ, C. 1985. Egyptian Magic, Aris and Phillips (Warminster).

KALENDER, W.A. 2000. CT: Fundamentals, System Technology, Image Quality, Applications, Publicis MCD Verlag (Munich).

KESSLER, D. and NUREDDIN A.H. 2005. 'Tuna-el-Gebel: Millions of Ibises and Other Animals' in S. Ikram (ed) Divine Creatures: Animal Mummies in Ancient Egypt, American University in Cairo Press (Cairo) 120-163.

LEEK, F.F. 1976. 'An ancient Egyptian mummified fish', Journal of Egyptian Archaeology 62, 131-133.

LEITH ADAMS, A. 1864. 'Notes on the Mummified Bodies of the Ibis and other Birds found in Egypt', The Edinburgh New Philosophical Journal XIX (11): 173-183.

LUCAS, A. 1962. Ancient Egyptian Materials and Industries, Fourth Edition revised by J.R. Harris, Edward Arnold Ltd (London).

MALEK, J. 1993. The Cat in Ancient Egypt, British Museum Press (London).

MOND, R. and MYERS O.H. 1934. The Bucheum, Egypt Exploration Society (London).

MOODIE, R.L. 1931. Roentgenologic Studies of Egyptian and Peruvian Mummies, Field Museum Press (Chicago).

NICHOLSON, P.T. 1994. 'Prelimiary report on the Sacred Animal Necropolis, north Saqqara', Journal of Egyptian Archaeology 80: 1-10.

NICHOLSON, P.T. 2005. 'The Sacred Animal Necropolis at North Saqqara: The Cults and Their Catacombs' in S. Ikram (ed) Divine Creatures: Animal Mummies in Ancient Egypt, American University in Cairo Press (Cairo) 44-71.

NOTMAN, D.N.H. 1986. 'Ancient Scannings: Computed Tomography of Egyptian Mummies' in A.R. David (ed) Science in Egyptology, University of Manchester Press (Manchester).

O'CONNOR, T. and O'CONNOR, S. 2005. 'Digitising and image-processing radiographs to enhance interpretation in avian palaeopathology', Documenta Archaeobiologiae 2005: 69-82.

OWEN, L.M. 2000. A radiographic study of thirty-nine animal mummies from ancient Egypt. Unpublished BSc dissertation, University of York.

OWEN, L.M. 2001. A radiographic investigation of the ancient Egyptian animal mummies from the Manchester Museum. Unpublished MSc dissertation, University of Manchester.

PAHL, W. 1986. 'Radiography of an Egyptian "Cat Mummy". An Example of the Decadence of the Animal Worship in the Late Dynasties?', OSSA Internationa Journal of Skeletal Research 12: 133-140.

PEDERSEN, J. 1997. 'Comparisons of vertebrae and otoliths measured directly and from radiographs', Fisheries Research 29: 177-282.

PEET, T.E. 1914. The Cemeteries of Abydos Part II – 1911-1912, The Egypt Exploration Fund (London).

PETTIGREW, T.J. 1834. History of Egyptian Mummies and an Account of the Worship and Embalming of the Sacred Animals, Longman, Rees, Orme, Brown, Green and Longman, (London).

PHILIPS, D.W. 1948. Ancient Egyptian Materials, The Metropolitan Museum of Art (New York).

QUIRKE, S. 2007. The Kahun Veterinary Papyrus. Personal communication regarding a translation of the Papyrus.

QUIRKE, S. and COLLIER, M. 2004. The UCL Lahun Papyri: religious, literary, legal, mathematical and medical, BAR Publishing (Oxford).

REITZ, E.J. and WING, E.S. 1999. Zooarchaeology, Cambridge University Press (Cambridge).

SELINCOURT, A. d. 1985. Herodotus: The Histories, Penguin (Middlesex).

SHACKLEY, M. 1985. Using Environmental Archaeology, B T Batsford Ltd (London).

SIEGEL, J. 1976. 'Animal Paleaopathology: Possibilities and Problems', Journal of Archaeological Science 3: 349-384.

SMITH, H.S. 1969. 'Animal domestication and animal cult in dynastic Egypt' in P.J. Ucko and G.W. Dimbleby (ed) The Domestication and Exploitation of Plants and Animals, Gerald Duckworth and Co. (London) 307-314.

SPENCER, A.J. 1982. Death in Ancient Egypt, Penguin (Middlesex).

TAYLOR, J.H. 2001. Death and the Afterlife in Ancient Egypt, British Museum Press (London).

TRUMBLE, K. 1996. Cat Mummies, Clarion Books (New York).

TYLDESLEY, J. 1999. The Mummy: Unwrap the Ancient Secrets of the Mummies, Carlton Books Ltd (London).

ZIVIE, A. and LICHTENBERG, R. 2005. 'The Cats of the Goddess Bastet'. in S. Ikram (ed) Divine Creatures: Animal Mummies in Ancient Egypt, American University in Cairo Press (Cairo) 106-119.

www.ingramcontent.com/pod-product-compliance
Lightning Source LLC
LaVergne TN
LVHW072257300926
849160LV00018B/1684